THE
SCIENCE OF
SCIENCE
FICTION

THE
SCIENCE OF
SCIENCE
FICTION

THE INFLUENCE OF
FILM AND FICTION ON
THE SCIENCE AND CULTURE
OF OUR TIMES

MARK BRAKE

Skyhorse Publishing

Copyright © 2018 by Mark Brake

All rights reserved. No part of this book may be reproduced in any manner without the express written consent of the publisher, except in the case of brief excerpts in critical reviews or articles. All inquiries should be addressed to Skyhorse Publishing, 307 West 36th Street, 11th Floor, New York, NY 10018.

Skyhorse Publishing books may be purchased in bulk at special discounts for sales promotion, corporate gifts, fund-raising, or educational purposes. Special editions can also be created to specifications. For details, contact the Special Sales Department, Skyhorse Publishing, 307 West 36th Street, 11th Floor, New York, NY 10018 or info@skyhorsepublishing.com.

Skyhorse® and Skyhorse Publishing® are registered trademarks of Skyhorse Publishing, Inc.®, a Delaware corporation.

Visit our website at www.skyhorsepublishing.com.

10 9 8 7 6 5 4 3 2 1

Library of Congress Cataloging-in-Publication Data is available on file.

Cover design by Sparth
Cover photographs by Sparth

Print ISBN: 978-1-5107-3936-9
Ebook ISBN: 978-1-5107-3937-6

Printed in the United States of America

For Zap

"Pythagoras with the looking glass reflects the full moon/
In blood, he's writing the lyrics of a brand-new tune" —Genesis

TABLE OF CONTENTS

INTRODUCTION

Here's a true story about my own science-fiction past. When I was a kid, I used to fill my days by reading lots of futuristic science-fiction stories. Stories about space, tales about time travel, entertaining yarns about gadgets and inventions, and comic-book accounts of the superpowers we might one day possess. The influence of these stories on my friends and me was simply huge. It occupied our waking hours, as well as our dreams. It gifted us fevered and fertile imaginations.

My friend Zap shared this fertile imagination, fed and fueled by science fiction. In fact, our imaginations became so fevered that, one day, Zap managed to convince me and some other friends that we could become superheroes. How would such superhero status be achieved? Simply by "necking [that's local slang for drinking] some superserum," made by Zap's own fair hand.

Yes, we must have been sweetly innocent (or stupid) to believe this fable and fabric of nonsense, straight out of the comic books, channeled through Zap's imagination, and back out into the "real" world. We waited and waited for the big day when our transformation from boy into superboy would come. We grew impatient. "But Zap, when are we going to become superheroes?" we justifiably asked.

"Don't worry about that, dudes. I've got it covered. I'm making some superserum in my dad's crib."

Yes. Superserum. In his dad's crib. On a housing project in Britain, back in those swinging sixties. Sure, Stan Lee's *Marvel* superhero tales in the 1960s were all about ordinary folk like us. Stan had revolutionized the genre with a large dose of reality: down-to-earth characters living everyday lives with stories peopled by folk who had personal problems. But superserum? In his dad's crib on a housing project?

It would have been wise to question the validity of Zap's claims, but wise we were not, as Yoda might chime. We were far too distracted by the promise of all the trappings superherodom brings: jumping over buildings, stopping speeding bullets, and of course, the girls. We got tired of waiting. Zap kept on promising, but nothing was delivered. In the end, we simply demanded the superserum. At once. With no delay. *Now!* We demanded the ability to jump over buildings, to stop speeding bullets, and of course, we demanded the girls. We followed Zap home and impatiently waited outside his house until he "brought forth" the superserum (in comics, inventors often appear to "bring forth" the invention).

As Zap triumphantly came out of the lab (his dad's crib), our breath was fully bated. What was he carrying, you may wonder, a sleek test tube full of some fluorescent and clearly cosmic superserum worthy of Hugo Strange? Perhaps a wizard potion in a cool Professor Snape-ish vial? Or a huge syringe with which to transform us forever? No, dear reader, Zap came out of his house with nothing more than a bowl full of ordinary water.

Gutted. In fact, mega-eviscerated. Picture this: my friends and I in a towering Hulk-like rage, reaching almost superhero proportions *without* the help of a superserum. After all that waiting, we were incensed. We were severely disappointed. We were *incandescent*! We hurled the bowl of water over Zap's head and stormed off into the sunset, disillusioned, but determined to find a brighter future.

A SCIENCE-FICTION WORLD

Maybe my early superserum experience is the main reason why, ever since, I've gotten a huge buzz on hearing that one of the predictions I read about as a kid has actually turned out to be true. If you've lived the dream from the start, and see the idea turn into reality, it can be simply inspiring.

It very much depends on perspective, though. Take this example. I was recently blown away by the fact that, sitting in a greasy British café, I was still able to watch a live soccer match on my phone. I was suitably enthusiastic to my kids. "Look, kids, I'm watching the football! It's the future. It's science fiction!" I said. My kids were "totes unfazed."

"No, dad, it's just a smartphone."

And yet, years before, I'd already found my niche as a professor of science communication at a British university where I'd invented study courses that focused on the future. One course was all about the search for alien life in space and another was all about the links between science and science fiction, but the one thing they had in common was our future on Earth and in space.

As a university science professor, I also did some writing for NASA and organized speaking tours for Russian cosmonauts, pioneers of space travel. Through my work, I found that humans have been making up science-fiction stories for hundreds of years, much longer than you'd think. Some of the earliest stories were about space voyages. They're in this book. They come from a time, the early 1600s, when astronomers had found that the Earth was in orbit around the Sun, and not the other way around. Ships had started voyaging around the globe four hundred years ago, so writers had already started to imagine sailing out into space.

Since then, there have been thousands of amazing science-fiction stories about aliens and time machines, spaceships and cyborgs, androids and the end of the world. These stories all have one thing in common: they're about the way science may affect our lives in the future, so it should be no surprise that this book is about the future.

We are the first generation to live in a science-fiction world. Media headlines declare this the age of automation. The TV talks about the coming revolution of the robot, tweets tell tales of jets that will ferry travelers to the edge of space, and social media reports that the first human to live for a thousand years has already been born. The science we do, the movies we watch, and the culture we consume is the stuff of fiction that became fact, the future imagined in our past, the future we now inhabit.

This book is the story of how science fiction shaped our world. No longer a subculture, science fiction has moved into the mainstream with the advent of the information age it helped realize. Explore how science fiction has driven science. This book will open your eyes to the way science fiction helped us dream of things to come, forced us to uncover the nature and limits of our own reality, and helped us build the science-fiction-driven world we live in today.

PART I

SPACE

It's where "no one can hear you scream" in Ridley Scott's *Alien*. It's the "final frontier" in *Star Trek*. It's the realm of Pandora in James Cameron's *Avatar*, along with a billion other habitable worlds of the imagination. The science fiction of space focuses on some facet of the natural world. Very often, the space theme is about that outward human urge to conquer and master the vast interstellar depths of the cosmos, in the classic manner of Captain Kirk and the legions of fleet and nimble ships of *Star Wars* and other space operas. In other tales, space is the vast, cold, and unsympathetic theater featured in *Alien*, or the immense void in *Interstellar*, a cosmos we may never come to terms with. The sheer indifference of space, such as featured in *The Martian*, reminds us that life may be frail and precious, but also tenacious, and that the universe is largely inhuman and deserted.

That Shakespeare of science fiction, H. G. Wells, really started something with *The War of the Worlds* back at the end of the nineteenth century. Wells's Martians were an early warning of what was to come. They were the first agents of the void, the first menace from space. The Martians were Wells's timely reminder that we may not be at the top of the universe's evolutionary ladder.

The alien is one of science fiction's greatest inventions. The alien is featured here in the space theme rather than monster theme for two reasons. For one, science fiction very often portrays the alien, such as the Na'vi in *Avatar* or the heptapods in *Arrival* as animated versions of nature. Second, the monster theme in science fiction focuses on the human condition, and, as we shall see, it's about the monster within us, not without.

A lot of science-fiction stories about space can be understood as a longing to escape our sense of being merely human. Earth is our prison. That's why we get tales in which the wonders and potential terrors of the universe are explored through the marvels of space travel, bringing tales of contact with extraterrestrial beings. Indeed, space stories have often been extrapolations of travelers' tales to lost or forgotten lands on Earth.

The space theme shows us that there are great similarities between science fiction and science, and that science fiction has been a huge influence on modern science and culture. Science fiction is an imaginative device for a kind of theoretical science: the exploration of imagined worlds. Scientists build models of hypothetical worlds and then test their theories. Albert Einstein was famous for this. His *Gedankenexperiment* or thought experiments led to his theory of special relativity, for example. The science-fiction writer also explores hypothetical worlds but with more scope. Scientists are meant to stay within bounded laws. Science fiction has no such boundaries. However, we can see that a spirit of "what if" is common to both science and science fiction.

There are many examples where science fiction has proposed theories far too speculative for the science of the day, but that have later proved to be prophetic. The theme of space contains some great examples—space travel, exoplanets, men on the moon, to name just a few. However, it's important to remember that the correctness of the science is not as important as its poetry or the sense of wonder and adventure experienced in pursuit of the science itself.

One of the best examples of the symbiosis of science and science fiction to be found anywhere is the question of life in space, or astrobiology. Through most of its history, science fiction has held a positive view of the possibility of alien life. Much of hard science fiction is made up mostly of two main sciences, physics and biology. Historically, physics came a lot earlier. Nicholas Copernicus shifted the center of the universe from the Earth to the Sun way back in 1543. Just over a century later, Isaac Newton produced a "system of the world," one of the first attempts to produce a theory of everything. As a result, early fictional accounts of alien life, from Johannes Kepler's *Somnium* (1634) to H. G. Wells's *The War of the Worlds* (1898), were based mostly on the physics in question. Simply put, the argument was based on the idea that since there are billions of stars in the universe, there must be millions, if not billions of planets. Plenty of places for ET to live. This line of reasoning is much like the principle of plenitude: everything that can happen *will* happen, and in a universe fit for life, many planets will bear bug-eyed monsters. For a long time, the mere physics of the matter has been fiction's main concern. Biology didn't come into it. The fact of the sheer number of stars and orbiting planets was enough to suggest that life on other Earths lay waiting in the vastness of deep space. The influence was fed back into science so that by the twentieth century, an entire generation of scientists was cast under the same spell, and huge investments were made in serious searches for alien life.

Meanwhile, a unified theory of biology didn't see the light of day until the second half of the twentieth century. It brought a modern synthesis of aspects of biology and was accepted by the great majority of working biologists. This evolutionary synthesis provided a new perspective on the question of ET. Biologists are far more skeptical about the possibility of complex alien life, let alone intelligence. We may, they concluded, be alone in the Universe after all, so let's venture into the space theme of science fiction to meet lost worlds and quantum universes, other Earths and the human future in the cosmos.

GUARDIANS OF THE GALAXY:
IS SPACE FULL OF EXTRATERRESTRIALS?

"Behind every man now alive stand thirty ghosts, for that is the ratio by which the dead outnumber the living. Since the dawn of time, roughly a hundred billion human beings have walked the planet Earth. Now this is an interesting number, for by a curious coincidence there are approximately a hundred billion stars in our local universe, the Milky Way. So for every man who has ever lived, in this Universe there shines a star. But every one of those stars is a sun, often far more brilliant and glorious than the small, nearby star we call the Sun. And many—perhaps most—of those alien suns have planets circling them. So almost certainly there is enough land in the sky to give every member of the human species, back to the first ape-man, his own private, world-sized heaven—or hell. How many of those potential heavens and hells are now inhabited, and by what manner of creatures, we have no way of guessing; the very nearest is a million times farther away than Mars or Venus, those still remote goals of the next generation. But the barriers of distance are crumbling; one day we shall meet our equals, or our masters, among the stars. Men have been slow to face this prospect; some still hope that it may never become a reality. Increasing numbers, however, are asking: 'Why have such meetings not occurred already, since we ourselves are about to venture into space?' Why not, indeed? Here is one possible answer to that very reasonable question. But please remember: this is only a work of fiction. The truth, as always, will be far stranger."

—Arthur C. Clarke, *2001: A Space Odyssey* (1968)

[Yondu is floating in the air, hanging on his arrow]
PETER QUILL: You look like Mary Poppins.
YONDU: Is he cool?
PETER QUILL: Hell yeah, he's cool.
YONDU: I'm Mary Poppins, y'all!
—James Gunn, *Guardians of the Galaxy Vol. 2 screenplay* (2017)

The idea of a universe full of extraterrestrial life has produced many of the best movie taglines: "In space, no one can hear you scream," "we are not alone," "a long time ago, in a galaxy far, far away," "the truth is out there," and, of course, "you only get one chance to save the galaxy twice." Science-fiction writers and directors have thought long and hard about the portrayal of creatures from other worlds. The predatory and possessive mother in Ridley Scott's *Alien*, the swirling and sentient sea in Steven Soderbergh's *Solaris*, and the wise, benevolent sage with curious sentence construction in George Lucas's *Star Wars* series have three very contrasting extraterrestrial types: alien as highly evolved killer, alien as ocean-planet, and alien as guiding mentor, perhaps only lacking a little gravitas due to being a puppet.

There are huge numbers of extraterrestrial races in the Marvel Comics Universe. So much so that the Guardians of the Galaxy, a band of former intergalactic outlaws, have teamed up to protect the Galaxy from planetary threats. The Galactic Council is an assembly of leaders of different alien empires from across that Universe with major races such as the Kree and the Skrulls presiding over scores of secondary alien races, mostly humanoid, but occasionally weird, such as the A'askvarii, a green-skinned race with octopus traits, three-toed taloned feet, three tentacles sprouting from each shoulder rather than arms, and closely spaced needle-like teeth, which would make a mess of an intergalactic burger. But, just in case you thought this was all new, take a look at how long aliens have dwelled in the human imagination.

Science fiction, driven by the discoveries of science, has been conjuring up extraterrestrials for many a moon. In fact, for much longer than you'd first imagine. Take the relationship between Italian astronomer Galileo and German math genius Johannes Kepler, for example. Moved by Galileo's discoveries with the telescope, Kepler was one of the first writers to imagine alien life. (And this is in the first few years of the 1600s!) Kepler made sure that the extraterrestrials stalking the characters in his protoscience fictional book, *Somnium*, published in 1634, were not humans. Instead, they are serpent-like creatures that are fit to survive their lunar, but quite alien, haunt. So, more than two centuries before Darwin, Johannes Kepler had been the first to grasp the bond between life forms and habitat, science and science fiction. But, generally speaking, before science fiction really rocketed into the creative imagination in the late nineteenth century, extraterrestrials were not normally portrayed as genuine alien beings. They were merely seen as humans and animals living on other worlds.

Charles Darwin changed all that, for Darwin essentially invented the alien.

Darwin's theory of evolution gave science fiction grounds for imagining what kind of life might evolve *off* Earth, as well as *on* it. From Darwin on, the notion of life beyond our home planet was linked with the physical and mental characteristics of the true extraterrestrial, and the idea of the alien became deeply embedded in the public imagination, so it's no surprise that the most credible extraterrestrials occur after his work. The archetypal alien, with its strange physiology and intellect, also owes much to H.G. Wells's first major take on Darwin: Wells's 1898 Martian invasion novel, *The War of the Worlds*. Wells's Martians are agents of the void. They are the brutal natural force of evolution, and history's first menace from space. Wells's genocidal invaders, would-be colonists of planet Earth, were so influential that the alien as monster became something of a cliché in the twentieth century—and the idea thrills us still. The alien as monster stalks the Nostromo in Ridley Scott's electrifying movie *Alien*, lies at the heart of each Dalek in *Doctor Who*, and briefly, to the soundtrack of ELO's *Mister Blue Sky*, consumes Drax during the surreal opening title sequence of *Guardians of the Galaxy Vol. 2*.

With advances in science, especially biology, early writers became more imaginative about alien life forms. Evolution traveled into space with the writings of French astronomer Camille Flammarion in 1872, barely a dozen years or so after Darwin published *The Origin of Species*.

Flammarion's three *Stories of Infinity* were ingenious tales of an intangible alien life-force. If natural selection was universal, there was no reason on Earth why the random process of evolution should merely produce humanoids on other planets. Distinguished British astronomer Fred Hoyle used his science to inspire his stories, but his fiction was not forced by his physics. Hoyle's first novel, *The Black Cloud* (1957), is about a living cloud of interstellar matter!

Polish science-fiction writer Stanislaw Lem pushed the creative imagination about alien life even further. In his famous *Solaris* (written in 1961 followed by movie adaptations in 1972 and 2002), now an entire planet enclosed by an ocean, Solaris is portrayed as a single organism with a vast yet strange intelligence that humans strive to understand.

And then, of course, there's the depiction of the extraterrestrial as wise, benevolent teacher, here to save us from ourselves. These are the kind of aliens that show up in films such as *Close Encounters of the Third Kind* (1977), where the extraterrestrials are presented as civilized and munificent aliens of superior intelligence. In the same way, aliens such as Yoda possess an almost saintly wisdom, and Marvel's Watchers are cast in a similar vein; one of the oldest species in the Universe, the Watchers are committed to watching and compiling knowledge on all aspects of life in the cosmos.

But, here's the point about all of these different portrayals of extraterrestrial life: even though science has made tremendous advances in the understanding of space during the twentieth and twenty-first centuries, scientists still have relatively little to say about the psychology and physiology of the alien. That's the job of science fiction, which has been conducting a kind of continuous thought experiment on the matter for centuries.

British science-fiction writer Arthur C. Clarke knows all about this relationship. He stressed the influence of science fiction on the alien life debate when he said in 1968, "I have little doubt that the Universe is teeming with life. The Search for Extraterrestrial Intelligence (SETI) is now a fully accepted department of astronomy. The fact that it is still a science without a subject should be neither surprising nor disappointing. It is only within half a human lifetime that we have possessed the technology to listen to the stars."

Clarke was very aware of the huge inspiration that science takes from fiction. In the history of the scientific debate on alien life, there have typically been two camps: physicists and biologists. The physical scientists such as astronomers have often tended toward a deterministic view of the possibility of extraterrestrial life. They focus on the physical forces in the Universe and make arguments based on the sheer number of stars and orbiting planets, which they feel is somehow statistically sufficient to suggest that other Earths lie waiting in the vastness of deep space. Fiction, for many centuries, led from the front with this same argument. And since Copernicus came before Darwin, and physics before biology, fictional accounts of alien life have usually been placed firmly in the pro-SETI, prolife camp of the alien debate. By the twentieth century, an entire generation of future SETI-hunters was cast under the same spell, and the imaginative power of science fiction meant that a huge investment of time and money was put into the serious scientific search for extraterrestrials.

But, as the twentieth century progressed, the story changed. Some scientists thought we might, after all, be alone in the Universe. In particular, biologists began to emphasize that while physics and fiction still think along deterministic lines, evolutionists are impressed by the incredible improbability of intelligent life ever to have evolved, even on Earth. Or, to put it in the powerful words of American anthropologist Loren Eisley,

"So deep is the conviction that there must be life out there beyond the dark, one thinks that if they are more advanced than ourselves they may come across space at any moment, perhaps in our generation. Later, contemplating the infinity of time, one wonders if perchance their messages came long ago, hurtling into the swamp muck of the steaming coal forests, the bright projectile clambered over by hissing reptiles, and the delicate instruments running mindlessly down with no report . . . in the nature of life and in the principles of evolution we have had our answer. Of men elsewhere, and beyond, there will be none forever."

It may be tomorrow or a decade or century from now until we discover if *Guardians of the Galaxy* is right. The day may come when we make the most shattering discovery of all time: the discovery of a thriving extraterrestrial civilization. When our current century dawned, we'd been imagining alien life for almost two and a half millennia, but as space agencies build flotillas of space telescopes to search for life in this unearthly Universe, the crucial questions remain unanswered.

The American space observatory *Kepler*, launched in 2009 to find Earth-like planets orbiting other stars, took off four hundred years after Galileo's first use of the telescope and is of course named after that first great Copernican theorist, Johannes Kepler. Based on *Kepler*'s early findings, Seth Shostak, senior astronomer at the SETI institute, estimated that "within a thousand light-years of Earth" there are "at least 30,000 habitable planets." Based on the same findings, the *Kepler* team projected "at least 50 billion planets in the Milky Way" of which "at least 500 million" are in the habitable zone. NASA's Jet Propulsion Laboratory was of a similar opinion. JPL reported an expectation of two billion "Earth analogues" in our Galaxy and noted there are around "50 billion other galaxies" potentially bearing around one sextillion Earth analog planets.

Over those last two and a half thousand years, a stunning array of writers and scholars, philosophers and filmmakers have devoted their energies to imagining life beyond this Earth. Their task has been to try reducing the gap between the new worlds uncovered by science and exploration and the fantastic strange worlds of the imagination. Their huge contribution has been important not only in the way that the fictional imagination has helped us visualize the unknown, but also for the way in which it has helped us define our place in a changing cosmos.

Stories rule. In the rich evolution of the question of extraterrestrial life, science fiction has influenced issues and debates in science, and in turn, popular culture has been inspired by scientific discovery and invention. The history of the alien has hinted at the revolutionary effects on human science, society, and culture that knowledge of another civilization will bring. If we may be so bold as to suggest that humanity is at least one way in which the cosmos can know itself, what more is out there to be discovered?

ARRIVAL:

HOW DO HUMANS COMPARE WITH ALIENS?

"If aliens visit us, the outcome would be much as when Columbus landed in America, which didn't turn out well for the Native Americans."

—Stephen Hawking, *The Guardian* (2010)

"Why should a vastly superior race bother to harm or destroy us? If an intelligent ant suddenly traced a message in the sand at my feet reading, "I am sentient; let's talk things over," I doubt very much that I would rush to grind him under my heel. Even if they weren't super-intelligent, though, but merely more advanced than mankind, I would tend to lean more toward the benevolence, or at least indifference, theory. Since it's most unlikely that we would be visited from within our own solar system, any society capable of traversing light-years of space would have to have an extremely high degree of control over matter and energy."

—Stanley Kubrick, *Playboy* (1968)

"The division of intelligent life into two categories—natural and artificial—may eventually prove to be meaningless. We may anticipate the synthesis of body parts . . . artificial intelligent beings of the future may be very long-lived. Their civilizations might be vastly longer-lived than civilizations like our own. Such civilizations could be very advantageous for interstellar contact among advanced communities . . . [and] might be able to transmit the treasures of science and the heritage of culture of a dead civilization into the cosmos for hundreds of millions of years."

—Carl Sagan and I. S. Shklovskii,
Intelligent Life in the Universe (1966)

Ian Donnelly: [narrating] "Why did they park where they did? The world's most decorated experts can't crack that one. The most plausible theory is that they chose places on earth with the lowest incidence of lightning strikes. But there are exceptions. The next most plausible theory is that Sheena Easton had a hit song at each of these sites in 1980. So, we just don't know."

—Eric Heisserer, *Arrival* (2016)

THE GREAT CHAIN OF BEING

Among the questions raised by the 2016 science-fiction movie *Arrival* is: where do humans stand in the cosmic scheme of things? In other words, how do we measure up against whatever extraterrestrial life might lie in wait beyond the Earth? This question has a considerable backstory. Just as the typical medieval town was walled in, so was the medieval universe: a walled-in cosmos, bounded between Heaven and Earth, closed to the ravages of change and time. Continuity was key. It was a connected cosmos, emanating from the realm of the Godhead in those old days of Faith, through the nested concentric spheres of crystalline perfection, which carried the planets on their courses and down to the dark and lowly corruptible Earth at its center.

This connected cosmos also included God's plan for all living things in the universe. And that plan was an ornate pageant of divine life known as the Great Chain of Being. On the Chain sat a cornucopia of creation, an infinite procession of links, stretching from God down to the lowliest form of life.

This scale of being, or *scala naturae*, was a strict hierarchical system, ranging from the highest perfection of the unchanging Spirit, who sat at the top of the chain, down to the fallibility of flesh at the core—mutable and corruptible Man.

It was a natural order, of sorts. Elements that sat at the foot of the chain, such as rock from which the Earth is made, merely existed. Moving up the chain, each successive link enjoyed more positive attributes to those below. Plants, then, possessed life, as well as existence; animals enjoyed the additional qualities of motion and appetite. Every imaginable creature and object had a place in this great scheme of things, but each position was determined in a rather anthropocentric way, often according to its usefulness to man. Wild beasts were superior to domestic ones, since they resisted training. Useful creatures, such as horses and dogs, were better than docile ones, such as sheep. Easily taught birds of prey were superior to lowlier birds, such as pigeons. Edible fish were higher up the totem pole than more dubious and inedible sea creatures.

Even aesthetics came into it. Attractive creatures such as dragonflies and ladybugs were considered worthier of God's glory than unpleasant insects such as flies and, no doubt, dung beetles. The poor snake languished at the very bottom of the animal segment, relegated as punishment for the serpent's alleged actions in the Garden of Eden. Some aspects of the Chain persist in popular culture today: the lion is still considered the king of all creatures, the oak the king of all plants, the eagle king of the sky.

It was not just unthinkable to abandon your place in the great continuous Chain; it was impossible.

This strict sense of permanence was crucial to the conception of the medieval world, and it was easy to see how this system became a useful means of control in feudal society. The Great Chain was used to justify the doctrine of the Divine Right of Kings, the idea that the monarch was subject to no earthly authority. The Chain had its reflection in this authoritarian social order that saw kings at the pinnacle, aristocratic lords below, and the great mass of peasants way down the societal scale. The king derived his right to rule directly from above and was not subject to the will of his people, the aristocracy, or any other estate of the realm, but to the will of God.

The importance of the Chain to early ideas of life in the universe is nicely summed up in the following quote, often referenced in texts of the Middle Ages, but written earlier by fifth-century philosopher Macrobius, one of the last pagan writers of Ancient Rome:

"Since, from the Supreme God Mind arises, and from Mind, Soul, and since this in turn creates all subsequent things and fills them all with life . . . and since all things follow in continuous succession, degenerating in sequence to the very bottom of the series, the attentive observer will discover a connection of parts, from the Supreme God down to the last dregs of things, mutually linked together and without a break. And this is Homer's golden chain, which God, he says, bade hang down from Heaven to Earth."

And this is where medieval man found himself. Spiritually speaking, man was a special case in the Chain. He was both mortal in flesh and potentially pure in spirit. The resolution of this struggle between his two aspects meant that he could either go the higher noble way of the spirit or be dragged down to the devil, the way of all flesh, just as Lucifer himself had fallen.

KINGS OF THE COSMOS

Now, you're probably wondering at this point, what on earth has all this to do with the movie *Arrival* and the human relationship with aliens? The answer, in short, is everything. From the very early days of science-fiction stories about alien life, the Great Chain of Being became the backstory of choice. The question had clicked with Johannes Kepler, author of one of the very first science-fiction stories in *Somnium*, almost as soon as Galileo discovered our Moon might be livable. And, centuries later, when Darwin gifted science the principle of evolution, a principle that might reign in heaven as well as on Earth, science fiction was again alive to the question of how man might measure up. Using Darwin as his guide, H. G. Wells quoted Kepler at the very start of *The War of the Worlds* Martian invasion: "But who shall dwell in these worlds if they be inhabited? Are we, or they, Lords of the World?" In other words: who's boss of the universe, and king of the cosmos? Man had assumed his dominion over nature, but how would we fare in that great expanse beyond this Earth? H. G. Wells had little doubt in the answer: man wouldn't measure up at all, so Wells created the myth of a technologically superior extraterrestrial intelligence. *The War of the Worlds* features the "men" of the future in alien form. They are what we may one day become. They are the tyranny of intellect alone. And imperial Britain, for once, is on the receiving end of interplanetary Darwinism. Look how brilliantly Wells frames the entire cosmic chain-of-being argument in one of the opening sections to *The War of the Worlds*:

"No one would have believed in the last years of the nineteenth century that this world was being watched keenly and closely by intelligences greater than man's and yet as mortal as his own; that as men busied themselves about their various concerns they were scrutinized and studied, perhaps almost as narrowly as a man with a microscope might scrutinize the transient creatures that swarm and multiply in a drop of water. With infinite complacency men went to and fro over this globe about their little affairs, serene in their assurance of their empire over matter. It is possible that the infusoria under the microscope do the same. No one gave a thought to the older worlds of space as sources of human danger or thought of them only to dismiss the idea of life upon them as impossible or improbable. It is curious to recall some of the mental habits of those departed days. At most terrestrial men fancied there might be other men upon Mars, perhaps inferior to themselves and ready to welcome a missionary enterprise. Yet across the gulf of space, minds that are to our minds as ours are to those of the beasts that perish, intellects vast and cool and unsympathetic, regarded this earth with envious eyes, and slowly and surely drew their plans against us. And early in the twentieth century came the great disillusionment."

Wells destroys the idea that man is the pinnacle of evolution, though (spoiler) the Martians turned out to be as complacent as man; earthly microbes eventually lead to their downfall. After Wells, the topic of alien contact and man's place on the cosmic evolutionary ladder became a twentieth-century obsession.

Science didn't even begin to come to terms with the potential shock of discovering an extraterrestrial intelligence until the 1960s, with Carl Sagan and I.S. Shklovskii's *Intelligent Life in the Universe* (1966), but decades before that, science-fiction writers like Olaf Stapledon and Arthur C. Clarke were preparing the public for the close encounters of alien contact. A close encounter of the first kind was a visual sighting of an unidentified flying object apparently less than 500 feet away. A close encounter of the second kind was a UFO incident in which a physical effect occurs, such as electronic interference, paralysis, or discomfort in the witness, or impressions in the earth, such as scorched vegetation or a chemical trace. Science fiction was far more interested in encounters of the third and fourth kind. The third kind being an encounter where beings are present, whether they be humanoid or robotic pilots of the craft, and the fourth kind involving some kind of human abduction.

This is where the movie *Arrival* comes into play. The Oscar-nominated blockbuster wondered what a message from a being with an entirely alien biology would look like. In a cosmic sense, humans are relatively newborn, and we've only been scanning our skies for alien craft or signals for the length of a human life. If a signal or craft is heading our way, it will likely be from a civilization that's at least a few hundred thousand years further evolved than us, higher up that great chain of being that science fiction has always been preparing us for. Would we really be able to understand what they had to say? It's likely that, even when the kings of the cosmos dumbed down their message, the secrets of what they had to say would remain a mystery.

HOW DID SCIENCE FICTION
PUT MEN ON THE MOON?

"For seven days and seven nights we sailed the air, and on the eighth day we saw a great country in it, resembling an island, bright and round and shining with a great light. Running in there and anchoring, we went ashore, and on investigating found that the land was inhabited and cultivated. By day nothing was in sight from the place, but as night came on we began to see many other islands hard by, some larger, some smaller, and they were like fire in color. We also saw another country below, with cities in it and rivers and seas and forests and mountains. This we inferred to be our own world.

—Lucian, *A True Story* (second century AD)

. . . the surface of the Moon is not perfectly smooth, free from inequalities and exactly spherical, as a large school of philosophers considers with regard to the Moon and the other heavenly bodies, but that, on the contrary, it is full of irregularities, uneven, full of hollows and protuberances, just like the surface of the Earth itself, which is varied everywhere by lofty mountains and deep valleys.

—Galileo Galilei, *The Starry Messenger* (1610)

"There will certainly be no lack of human pioneers when we have mastered the art of flight. Who would have thought that navigation across the vast ocean is less dangerous and quieter than in the narrow, threatening gulfs of the Adriatic, or the Baltic, or the British straits? Let us create vessels and sails adjusted to the heavenly ether, and there will be plenty of people unafraid of the empty wastes. In the meantime, we shall prepare, for the brave sky-travelers, maps of the celestial bodies—I shall do it for the Moon, you Galileo, for Jupiter.

—Johannes Kepler,
Conversation with the Starry Messenger (1611)

"I foresaw very well, that the vacuity . . . would, to fill up the space, attract a great abundance of air, whereby my box would be carried up; and that proportionable as I mounted, the rushing wind that should force it through the hole, could not rise to the roof, but that furiously penetrating the machine, it must needs force it upon high.

—Cyrano de Bergerac,
The States and Empires of the Moon (1657)

To answer this question, we could start with Italian astronomer Galileo Galilei. Four hundred years ago, Galileo famously wielded the newly invented telescope like a weapon of discovery. A new universe was unveiled, and the Moon became a vital piece in a new cosmic puzzle. Galileo's revolutionary pamphlet, *The Starry Messenger*, told the tale of his discoveries with the telescope. Written in 1610, it urged the reader to imagine walking on the lunar mountains and craters, just like here on our home planet. It was the first time the Moon became a real object for the great majority of people. Before Galileo, the Moon for many was just a disc in the sky, but with the telescope it became an object of wonder, and we began to contemplate the possibilities: Was there life on the Moon, and how might humans one day walk those craggy craters?

And yet, science fiction had already sent men to the Moon. Years before Galileo wielded his spyglass, German math scholar Johannes Kepler had begun imagining journeys to Earth's natural satellite (indeed, the word satellite was actually coined by Kepler). Kepler's book, *Somnium*, though published in 1634, had first germinated in Kepler's mind as early as 1593. It was one of the first fictional Moon voyages with a strong scientific flavor, in which Kepler imagined an alien life fit for a lunar landscape.

The Scientific Revolution itself began with Galileo's discoveries with the telescope. The countless stars and impure spots on an allegedly perfect Sun that could only be seen with the aid of a spyglass were new evidence to the lie of the perfect and immutable heavens that old Aristotle claimed almost two thousand years earlier. Still, it was mostly Galileo's discovery of other worlds that sparked a revolution: four entirely new worlds in the shape of the main moons in orbit around Jupiter, a focus of gravitation other than the Earth, and the discovery of a "world" in the Moon with its terrestrial mountains and craters.

Both Kepler and Galileo produced a map of the knowable, just as the unknown was at the point of becoming known. Galileo's *Starry Messenger* implied that Kepler's creatures may well be dwelling on the Moon. It was a vital new piece of evidence in the debate on the existence of extraterrestrial life. Yes, it's the first time the Moon became a real object for us, but at that same instant, we also felt a sense of wonder or estrangement from this new reality. Estrangement implies a state of imperfect knowledge. It is the result of coming to understand what is just within our mental horizons.

THE BIRTH OF SCIENCE FICTION

Like science, science fiction plays on that same sense of wonder. Galileo's spyglass discoveries of Jupiter's moons also led Kepler to wonder if they too may be inhabited. Galileo had to assume that shadows on the Moon have similar causes to shadows on Earth in order to understand the Moon's difference from Earth, yet so great a scholar as Kepler evidently needed to believe in aliens in order to render Galileo's discovery thinkable. Kepler realized that to understand the Moon it was not enough to put one's observations into words. The words themselves had to be transformed by a new sort of fiction. That's why there is something revolutionary about *Somnium* in the history of science. Throwing words at the Moon, as it were, has a dialectic effect—the words come back to us changed. By imagining the strange worlds of science fiction, we come to see our own conditions of life in a new perspective.

Science fiction starts with the paradigm shift of the Scientific Revolution, the shift of the old universe into the new. The universe of our ancestors had been small, static, and Earth-centered. It had the stamp of humanity about it. Constellations bore the names of Earthly myths and legends, and a magnificence that gave evidence of God's glory. The new universe of Kepler and Galileo was decentralized, inhuman, infinite, and alien. Galileo's discoveries had just about proved that the universe was not geocentric, but the revolution cut another way—it made Earth an alien planet. After all, if the Earth was not central, then neither was humanity. Earth may not be the only inhabited planet, and so it became the job of science fiction to come to terms with the cultural shock created by finding our marginal position in a universe fundamentally inhospitable to us, to make human what is alien, and to put that stamp of humanity back on the cosmos. It's what science fiction has been doing ever since.

When it comes to the Moon, the influence of science fiction is astounding. When Kepler first heard of Galileo's observations through the new telescope, he must have had Leonardo da Vinci in mind. Kepler wrote to Galileo, saying he felt there would be no lack of human pioneers when man had mastered the art of flight, so with a leap of the creative imagination, Kepler invented the spaceship! He talked about creating craft "adjusted to the heavenly ether" for the "brave sky-travelers" who are "unafraid of the empty wastes."

Kepler's journey to the Moon was one of the first-ever works of science fiction. Incredibly, the Greek scholar Lucian had also written about a Moon voyage more than 500 years earlier. The striking difference was this. Lucian's *A True Story* is fantasy, while Kepler's *Somnium* was a conscious effort to understand the new physics of the Scientific Revolution—what would the cosmos look like from another world? The key factor in developing Kepler's consciousness was the evidence provided by the newly invented telescope.

Progress in science inspired further creativity from science fiction. In 1638, hot on the heels of Kepler's work came *The Man in the Moone*, written by Francis Godwin, the Bishop of Llandaff in Wales. Godwin's volume was equally extraordinary. Again, it first gelled in Godwin's imagination in the 1590s. The book explored the possibility of a space voyage to another world, and get this: Godwin's is the first English book in history to portray alien contact. *The Man in the Moone* captured the imagination of John Wilkins, First Secretary of the Royal Society. Wilkins's own work was revised to take account of the popularity of Godwin's book, and the notion that it was just a matter of time before a lunar encounter took place. Wilkins proposed a flying machine would one day wing its way moonward.

Both Kepler and Wilkins made remarkably early predictions of vessels equipped for a lunar voyage. Cyrano de Bergerac followed suit in spectacular style. Cyrano was a notorious French duelist, satirist, and freethinker whose life is immortalized by many romantic legends, but one of his innovations is lesser known. According to Arthur C. Clarke, Cyrano's *The States and Empires of the Moon* (1657) is to be credited for conceiving the ramjet, a form of jet engine that contains no moving parts, so not only did medieval science fiction dream up spacecraft, it also dreamt up a way of propelling vessels to the Moon.

The rest, as they say, is history. By the time science had made man the machine age of the nineteenth century, the idea of lunar life no longer held any credibility. The Moon was dead. And yet, at just over a mere light-second away, the lunar landscape was there to be conquered and claimed for science. For pulp-fiction writers in particular, reaching the Moon now became an article of faith. Foremost was American writer Robert A Heinlein. Books such as his *Rocket Ship Galileo* (1947) portrayed our satellite as a stepping-stone for the development of the solar system at large. Crucially, Heinlein's *The Man Who Sold the Moon* (1950) told a tale of the fight to finance the first Moon-shot, and how to sell the myth of space conquest to the world. Sound familiar? Fact followed fiction. Destination Moon became an obsession for Cold War politicians who saw the propagandist coup that a manned Moon landing signified. Apollo got there first, and a dozen US astronauts from various missions have been the only humans to set foot on lunar soil. So far.

SHOULD SCIENCE MAKE A REAL-LIFE JURASSIC WORLD?

"God created dinosaurs. God destroyed dinosaurs. God created Man. Man destroyed God. Man created dinosaurs. Dinosaurs eat man . . . woman inherits the earth."

—Michael Crichton, *Jurassic Park* (1990)

"But now science is the belief system that is hundreds of years old. And, like the medieval system before it, science is starting not to fit the world any more. Science has attained so much power that its practical limits begin to be apparent. Largely through science, billions of us live in one small world, densely packed and intercommunicating. But science cannot help us decide what to do with that world, or how to live. Science can make a nuclear reactor, but it cannot tell us not to build it. Science can make pesticide but cannot tell us not to use it. And our world starts to seem polluted in fundamental ways—air, and water, and land—because of ungovernable science."

—Michael Crichton, *Jurassic Park* (1990)

"Scientists are actually preoccupied with accomplishment. So, they are focused on whether they can do something. They never stop to ask if they should do something. They conveniently define such considerations as pointless. If they don't do it, someone else will. Discovery, they believe, is inevitable. So, they just try to do it first. That's the game in science. Even pure scientific discovery is an aggressive, penetrative act. It takes big equipment, and it literally changes the world afterward. Particle accelerators scar the land and leave radioactive by-products. Astronauts leave trash on the moon. There is always some proof that scientists were there, making their discoveries. Discovery is always a rape of the natural world. Always.

—Michael Crichton, *Jurassic Park* (1990)

"Open the door, get on the floor
Everybody walk the dinosaur"

—Was (Not Was), *Walk the Dinosaur* (1987)

LOST WORLDS

Until science fiction came along, lost worlds were unheard of. A typical "lost world" tale would first find our adventurer somewhere in the civilized world, usually London. Armed with a tall story or an ancient scroll, our hero sets off to unknown lands and lost civilizations to find secret powers of great antiquity. The quintessential lost world was Plato's Atlantis, but the classic lost-world stories of the Victorian age such as H. Rider Haggard's *King Solomon's Mines* and *She: A History of Adventure*—the famous "She Who Must Be Obeyed" —are with us still. Such stories survive in video-game adventures such as the *Tomb Raider* series with Lara Croft (she too must be obeyed), but lost-world tales found a new and scientific form in Michael Crichton's *Jurassic* series.

First, though, was the exploration and exploitation of the natural world. Before you can go find a lost world, after all, it's got to go missing in the first place. The medieval voyages of discovery had really opened up our planet to piracy and plunder. One of the main architects of the new philosophy of medieval science was an English statesman named Francis Bacon, the key prophet and publicist of the new age, a kind of Renaissance spin doctor. He believed that organized science would forge material progress and then seized on the idea that understanding nature was the main means of taming the planet for profit. He wasn't wrong.

In an echo of Michael Crichton's remarks in *Jurassic Park* about "ungovernable science," but many centuries before, Bacon foresaw a utopia and empire of science. In his diaries, he claimed to be seeking to "enlarge the bounds of human empire to make all things possible." Bacon believed nothing should get in the way of progress, and, to this end, he developed a flawed medievalist ideology of power with his idea of

"Monstrosity." There were peoples, Bacon said, who had degenerated from the laws of nature and become monstrous. Among such multitudes, who Bacon believed deserved destruction, were (according to Bacon's diaries) "West Indians, Canaanites, pirates, land rovers, assassins, Amazons, and Anabaptists." Armed with this kind of philosophy, the expropriating British Empire expanded, and native colonial peoples were dispossessed, shot, poisoned, and diseased.

Before the 1770s, large parts of the world remained unknown to Europeans. As science and technology grew, and as modern "civilization" crept around the globe, fantastic travelers' tales became very popular. One of the very first examples was Ludvig Holberg's *Nicolaii Klimii iter Subterraneum* (1745). Translated as *A Journey to the World Underground*, or simply *Niels Klim's Underground Travels*, the tale tells of a young Norwegian who stumbles down into the Earth to discover an inner planet populated by intelligent nonhuman life-forms. The novel starts when Niels Klim returns from the university in Copenhagen where he studied philosophy and theology. His scientific curiosity drives him to investigate a strange cave hole in the Earth on the mountain above the town. Niels ends up falling down the hole and after a short while finds himself floating in free space. The Earth, it seems, is hollow but populated by an array of alien, but intelligent creatures.

Perhaps a more familiar take on such a lost world story is Jules Verne's *Journey to the Centre of the Earth*, written in 1864. Verne's creative journey had begun in 1863 with the first of 63 *Voyages Extraordinaires: Voyages in Known and Unknown Worlds*. An early advertisement claimed Verne's goal was, "to outline all the geographical, geological, physical, and astronomical knowledge amassed by modern science and to recount, in an entertaining and picturesque format that is his own, the history of the universe." Some mission.

However, whereas *Journey to the Centre of the Earth* is classic Jules Verne science fiction, he was quite unapologetic about the penetrative thrust of science into nature. Verne's book is a voyage through a subterranean world, and a conquest of space. The novel's main idea is this: nature is a cypher to be cracked. It's a journey into the depths of evolutionary time. Once into the Earth's subterranean caverns, grottos, and waters,

Verne's explorers find the interior alive with prehistoric plant and animal life. A herd of mastodons, giant insects, and a deadly battle between an Ichthyosaurus and a Plesiosaurus follow. Verne's book promotes a giddy confidence in progress, and a predictable cosmos in which the unknown is easily assimilated into our taxonomies.

The sheer pace of dizzying progress got even worse in Victorian times, when the emergent sciences of biology and geology made the modern feeling of alienation even worse. Science fiction began to try to repair the separation from nature, to reload the emptiness, to somehow jack-in to the void. Verne's fiction finds lost worlds and dinosaurs by exploring geographical space. Michael Crichton's *Jurassic Park* did the same by jacking into our genetic past.

THE GENE GENIE

It's hardly surprising that we get excited by the prospect that, someday, dinosaurs could live again. Look at what genetics has done for human history. Gene markers in human blood are a type of time machine. Inside a drop of blood is the best history book ever written, with everyone on Earth carrying a unique chapter in their veins. We've learned how to read the time machine in our genes by taking blood samples from people all around the planet. The conclusions are stunning. All humans alive today are related by blood in one big family tree. It really wasn't that long ago that there were only about two thousand humans living on a single continent of Africa. Our blood tells the tale of a small group of ancient humans who left Africa on a long journey; those of us alive today are their children. We are still working out the story of how our ancestors came to populate the Earth in surely the most incredible journey in human history.

What if we could use genetics to recover lost worlds?

In *Jurassic Park*, Michael Crichton was ahead of the curve. To bring dinosaurs back from the dead would require an intact genome, notoriously hard to find. You'd need to find a genome template from an ancient sample with sufficient data to make more than just a few genes. Tricky. Also, the quality and quantity of prehistoric DNA would depend on the preservation and degradation of the sample's DNA. True, DNA is surely stable enough for nature to trust it through evolutionary time. Nonetheless, DNA *does* degrade over time, so the challenge would be finding enough intact pieces in prehistoric DNA, especially in amber-trapped insects found in South American rain forest habitats. There might not even be enough intact DNA to clone a single deadly raptor claw, and yet the point was that Crichton fueled our dreams. He cloned dinosaurs. He let the imagination of science soar and raised expectations of new and fantastical future discovery.

Recent advances in genetic tech have boosted the hope of being able to clone prehistoric creatures in the future. Scientists have successfully sequenced a 10,000-year-old woolly mammoth, a 38,000-year-old Neanderthal, and the 80,000-year-old genome of a young female from an early species of Homo sapiens called the Denisovans—a close relation to the Neanderthals. The project team even reported the Denisovan girl had brown skin, eyes, and hair. Finally, pushing back the barriers even further into our prehistoric past, the entire genetic sequence of a 700,000-year-old extinct species of horse was recently published in the journal *Nature*.

A final important aspect of Crichton's science fiction remains, and that's the question of chaos theory. It's encapsulated in the quote, "Scientists are actually preoccupied with accomplishment. So they are focused on whether they can do something. They never stop to ask if they should do something." The words come from Crichton's lead character, Dr. Ian Malcolm, the skeptical chaotician in the fiction and film. Malcolm goes on to say that "Science cannot help us decide what to do with that world, or how to live. Science can make a nuclear reactor, but it cannot tell us not to build it. Science can make pesticide but cannot tell us not to use it. And our world starts to seem polluted in fundamental ways—air, and water, and land—because of ungovernable science." One solution may be taking such decisions out of the hands of private capital, and into the public arena. To ensure that future science projects don't simply count down the seconds to disaster, we the people need to make the decisions.

INTERDIMENSIONAL RICK AND MORTY:
DO PARALLEL WORLDS EXIST?

"If a coin comes down heads, that means that the possibility of its coming down tails has collapsed. Until that moment the two possibilities were equal. But on another world, it does come down tails. And when that happens, the two worlds split apart."
—Philip Pullman, *Northern Lights* (1995)

"Philosophically, the universe has really never made things in ones. The Earth is special and everything else is different? No, we've got seven other planets. The Sun? No, the Sun is one of those dots in the night sky. The Milky Way? No, it's one of a hundred billion galaxies. And the universe—maybe it's countless other universes."
—Neil deGrasse Tyson, *Cosmos: A Spacetime Odyssey* (2014)

MORTY: "That, out there, that's my grave."
SUMMER: "Wait, what?"
MORTY: "On one of our adventures, Rick and I basically destroyed the whole world, so we bailed on that reality and we came to this one, because in this one, the world wasn't destroyed, and in this one, we were dead. So, we came here and we buried ourselves, and we took their place. And every morning, Summer, I eat breakfast twenty yards away from my own rotting corpse."
SUMMER: "So, you're not my brother?"
MORTY: "I'm better than your brother. I'm a version of your brother you can trust when he says "Don't run." Nobody exists

on purpose, nobody belongs anywhere, everybody's gonna die. Come watch TV."

—*Rixty Minutes,* written by Tom Kauffman and Justin Roiland, *Rick and Morty 1.08* (2014)

MORTY: "Rick, what about the reality we left behind?"
RICK: "What about the reality where Hitler cured cancer? Just don't think about it, Morty."

—*Rick Potion #9,* written by Justin Roiland, *Rick and Morty 1.06* (2014)

"I don't do magic, Morty, I do science. One takes brains, the other takes dark eyeliner."

—*Pickle Rick,* written by Jessica Gao, *Rick and Morty 3.03* (2017)

"Life could be horrible in the wrong trouser of time."
—Terry Pratchett, *Guards! Guards!* (1989)

QUANTUM UNIVERSES

Is this the best of all possible worlds? One would certainly hope not. The question was first lit in the nervous system of German philosopher Gottfried Leibniz. Gottfried was a mercurial polymath who not only invented a branch of math called calculus, independently of Isaac Newton, but also refined the binary number system, which today is the basis of almost all digital computers. No doubt flush with all this success, and clearly not a man to shirk the tricky questions, Gottfried turned his keen mind to try solving the problem of evil. His thinking went something like this: if God is good, omnipotent, and omniscient, how come there is so much suffering and injustice in the world? You must admit, Gottfried had a point. His solution in many ways preempted a science-fictional obsession that was to follow. He made God a kind of "optimizer." God simply chose from a host of all original possibilities, and since God is good, this world must be the best of all possible worlds. Convinced? No, me, neither. I'm sure there's a loophole in there somewhere. And the idea of God as optimizer makes him sound very much like Rick Sanchez. Rick is the genius alcoholic scientist, based on Dr. Emmett Brown from *Back to the Future*, whose reckless and nihilistic antics feature in the animated science-fiction television series, *Rick and Morty*. The pair divides their time between dull domestic drudgery and interdimensional adventures inspired by the findings of quantum mechanics.

Let's consider quantum mechanics.

Chief among the mysteries of quantum theory is the notion that our Universe is merely one of many. This is known as the "many-worlds interpretation." Propounded by the likes of American physicist John Wheeler, and popularized by science fiction writers the world over, the theory goes further. It imagines an infinite number of parallel worlds or Universes, making up a "multiverse," that together comprise all of physical reality. If you've seen *Rick and Morty*, then this is all beginning to sound very familiar.

Not only that, but such a multiverse is said to contain all possible earthly histories and physical universes. Head hurting? Don't worry; it's quite natural. In fact, quantum supremo John Wheeler once said, "If you are not completely confused by quantum mechanics, you do not understand it." This would be more comforting if it didn't sound *exactly* like the kind of thing Rick would say to Morty. Science fiction plays with quantum theory in many different forms. Parallel worlds may also be conjured in stories under names such as "other dimensions," "alternate universes," "quantum universes," or even "alternate realities."

The speculative idea that other worlds lie in parallel to ours is older in fiction than it is in the "fact" of quantum theory. The most distinguished example, of course, is Lewis Carroll's 1865 story, *Alice's Adventures in Wonderland*, whose heroine pops into a parallel reality via a rabbit hole. But even the *Alice* tale itself may not be all it seems. *Alice* was the brainchild of math Oxford don Charles Dodgson, who really wanted to satirize the new and abstract math that had begun to filter into Oxford in the 1860s. The story's inclusion of a hookah-smoking caterpillar, a meeting with a duchess whose baby turns into a pig, and a tea party with the Mad Hatter and the March Hare are all thought to be ridiculing contemporary mathematicians. Dodgson didn't like the way that math was becoming so extraordinarily abstract and pure and less to do with either symbolic logic, which was his particular subject, or the beauty of plain Euclidian geometry, which he loved. Magic mushrooms, babies morphing into pigs, and absurd questions such as "Why is a raven like a writing desk?" were all meant to show how pointless and wrong these new theories were. Perhaps the most ingenious satire is the inclusion, in later editions, of the story of the Cheshire Cat, who disappears leaving only a grin. It's Dodgson's humorous way of making a serious academic point about the futility of abstraction.

Another notable Victorian parallel-world story is Edwin Abbott's *Flatland* (1884). Abbott's reality, a world with an alternate physics from our own, describes a world with only two dimensions, rather than the usual three, assuming we ignore time as the fourth, as H.G. Wells had not yet dreamed it up!

SAIL ON, RICK AND MORTY

During the last century or so, science fiction has popularized the "many-worlds interpretation" of reality and made it a main feature of modern culture. A wonderful example of this is American writer Phillip Jose Farmer's 1952 novel, *Sail On! Sail On!* The book describes an alternate 1492 AD in which the Earth *is* flat. The parallel-world physics in *Sail On!* is so strange that Columbus sails over the edge of the world and into Earth orbit, never to return from his mission. This simple tweaking of physics makes you wonder what our world might look today if Columbus had been dumped into the dustbin of history and the likes of the Incas, Aztecs, and Native Americans were left alone to prosper.

Phillip Jose Farmer was something of a parallel-world mastermind. Another of his tales seems like a whimsical perspective on Leibniz's original idea. Farmer's *The Unreasoning Mask* (1981) features a multiverse, each universe contained within a different cell of the body of God. The only means of travel between alternative universes is through the cell walls, which is wounding to the growing body of the still-infant God. (Incidentally, *The Unreasoning Mask* is a little like an episode of Rick and Morty called *Anatomy Park*. A parody of *Jurassic Park*, the episode is a fantastic voyage through the body of a homeless man, which houses a number of deadly diseases that escape their enclosures.)

Most science fiction is content with a multiverse that contains one or two parallel worlds besides ours, but *Rick and Morty* births worlds without end. There's a world whose climate is so perfect that nature allows megatrees to be cultivated. Another world sells, at every drugstore corner, a superserum that fixes broken bones. And yet another world in which pizzas obsessively eat people. And that's just for starters.

Though the makers of *Rick and Morty* declare the show the "most scientifically accurate animated comedy" with tongue firmly imbedded in cheek, like Douglas Adams's *The Hitchhiker's Guide to the Galaxy* before it, the show's main purpose is parody. Witness the show's alternate history that sees Albert Einstein discovering relativity after being beaten up by testicle-headed fourth-dimensional beings. *Rick and Morty*'s novel spin on the multiverse takes "real" physics and plants it firmly in the public imagination. As Matthew Kleban, a cosmologist at New York University, says, "I think it's much better that people are interested enough in the science that they write stories about it, than if they didn't care."

Among the physicists themselves, there's no consensus on whether such parallel worlds, or the multiverse, actually exists. There isn't even an agreement on what those words mean. While some think the multiverse could exist and try to find traces of our universe colliding with a neighbor, others regard the theory as utter nonsense since it suggests all things are equally likely, and so nothing can negate the idea. Others argue that the main trouble with the "best of all possible worlds" approach, from Leibniz on, is that authors very often depict dystopias, thereby promoting passivity. Our world is far better, so why change things?

Cosmologists David Deutsch and Max Tegmark are among the theoreticians who believe that parallel worlds could actually exist. In an issue of *Scientific American*, Tegmark calculated that our Milky Way galaxy has a twin, in which there is a twin Earth, which in turn contains a twin of you. No doubt there's also a twin of this book. Come to think of it, if the "many-worlds interpretation" of reality is true, then in some parallel world, this book will win an *American Association for the Advancement of Science* book award—and also not win it.

THE WAR OF THE WORLDS:

HOW DID SCIENCE FICTION CONVINCE US ALIENS WOULD INVADE?

"'But who shall dwell in these worlds if they be inhabited? Are we or they Lords of the World? And how are all things made for man?' —Kepler (quoted in *The Anatomy of Melancholy*)"
　　　　　　　　　　　—H.G. Wells, *The War of the Worlds* (1898)

"We know now that in the early years of the twentieth century this world was being watched closely by intelligences greater than man's and yet as mortal as his own. With infinite complacence people went to and fro over the earth about their little affairs, serene in the assurance of their dominion over this small spinning fragment of solar driftwood which by chance or design man has inherited out of the dark mystery of Time and Space. Yet across an immense ethereal gulf, minds that to our minds as ours are to the beasts in the jungle, intellects vast, cool and unsympathetic, regarded this earth with envious eyes and slowly and surely drew their plans against us. In the thirty-ninth year of the twentieth century came the great disillusionment. On this particular evening, October 30, the Crosley service estimated that thirty-two million people were listening in on radios.
　　　　—Orson Welles, *The War of the Worlds* radio broadcast,
　　　　　　　　　　　　　　　　　　Sunday, October 30, 1938

Playboy: "You'll admit, though, that extraterrestrials are commonly portrayed in comic strips and cheap science-fiction films as bug-eyed monsters scuttling hungrily after curvaceous Earth maidens."

Kubrick: "This probably dates back to the pulp science-fiction magazines of the twenties and thirties and perhaps even to the Orson Welles Martian-invasion broadcast in 1938 and the resultant mass hysteria, which is always advanced in support of the hypothesis that contact would cause severe cultural shock."

—Stanley Kubrick, *Playboy* (1968)

"The important point is that all the standard attributes assigned to God in our history could equally well be the characteristics of biological entities who billions of years ago were at a stage of development similar to man's own and evolved into something as remote from man as man is remote from the primordial ooze from which he first emerged . . . in an infinite, eternal universe, the point is that anything is possible, and it's unlikely that we can even begin to scratch the surface of the full range of possibilities. But at a time when man is preparing to set foot on the Moon, I think it's necessary to open up our Earth-bound minds to such speculation. No one knows what's waiting for us in our universe. I think it was a prominent astronomer who wrote recently, 'Sometimes I think we are alone, and sometimes I think we're not. In either case, the idea is quite staggering.'"

—Stanley Kubrick, *Playboy* (1968)

THE SHAKESPEARE OF SCIENCE FICTION

In 1968, *Playboy* interviewed American film director Stanley Kubrick. The occasion, of course, was Kubrick's new film, *2001: A Space Odyssey*, conceived and developed with British science-fiction writer Arthur C. Clarke during the infamous extraterrestrial hypothesis, which peaked between 1966 and 1969. The hypothesis and the surrounding hysteria was based on the firm conviction that worldwide UFO sightings and live close encounters with visiting aliens were actually real, a hypothesis vastly influenced by science fiction itself.

Celebrated for the maturity of its portrayal of mysterious, existential, and elusive aliens, Kubrick and Clarke's *2001* had raised science-fiction cinema to a new level. Eminent film critic, Roger Ebert, when asked which films would remain familiar to audiences 200 years from now, selected *2001*. Another critic, quoted in Youngblood's *Expanding Cinema* in 1970, claimed the picture was an "epochal achievement of cinema" and "a technical masterpiece." The film, not the book, made Clarke the most popular science-fiction writer in the world. Kubrick's masterpiece, which made dramatic and sophisticated use of the extraterrestrial hypothesis, quickly became a classic discussed by many, if not understood by all. And yet when asked by *Playboy* why the idea of alien invasion was so popular, Kubrick took to traveling back in time, first to H.G. Wells, then to Orson Welles.

The impact of British science-fiction writer H.G. Wells was colossal. Author of classic works such as *The Time Machine* (1895) and *The War of The Worlds* (1898), Wells was responsible for igniting both the *space* and *time* themes in the genre of science fiction, and in the public imagination. Wells created the nexus of the alien, armed with its potential for probing human evolution. And as Leonard Isaacs says in his 1977 book, *Darwin to Double Helix: The Biological Theme in Science Fiction,* Wells's early books "are, in their degree, myths; and Mr. Wells is a myth-maker."

Kubrick was well aware of the fact that, once developed by Wells, the alien idea proved a potent way of exploring the singularity or insignificance of humanity. During such explorations, the question of the character of alien interaction became an issue, which later affected the extraterrestrial hypothesis and the SETI science program. As Brian Aldiss put it in his 1973 book about the history of science fiction, *Billion Year Spree*, "Wells is the Prospero of all the brave new worlds of the mind, and the Shakespeare of science fiction."

In his original 1898 *The War of the Worlds* novel, H.G. Wells had made his Martians agents of the black void of space. Their weird physiology and intellect made them the prototypical alien. The Martian Tripods tower over men physically, as the vast intellects of their occupants tower over human intelligence. Bodily frail, but mentally intense, the Martians and their superior machines are instruments of human oppression. Their weapons of heat rays and poison gas are dehumanizing devices of mass murder. All attempts at contact are futile, furthering the idea of the aliens as an unrelenting force of the void.

Wells made the Martians seem even more alien by saying nothing about their culture. It seems to have wasted away by some entropic decay and whittled down to nothing more than a cosmic justification to invade the Earth. Nor do the Martians have any interest in human culture. Like vampires, they are interested only in human blood—vast, cool, unsympathetic, and alien in tooth and claw, as it were. And if you feel at all sorry for the desperate Martians, it soon becomes clear you may merely be projecting emotion onto creatures that are fundamentally inhuman.

The Martians are also political. Wells evidently has the Martians brutally colonize Earth, but, as he says in *The War of the Worlds*, "before we judge them too harshly we must remember what ruthless and utter destruction our own species has wrought upon its own inferior races. Are we such apostles of mercy as to complain if the Martians warred in the same spirit?"

Finally, in writing about the alien, Wells is also writing about our own world. The Martians are a veiled criticism of the industrial age, with its application of science to industry. In this fashion, *The War of the Worlds* is a rage against the social machine of humans, our reduction to anonymous cattle, the indifference at any attempts to communicate the inhumanity of the system, and the feeling of alienation. The whole scenario, taken together, inspires readers' fear and loathing of the Martians, and their alien nature.

MARTIANS ON THE RADIO

The radio version of *The War of the Worlds* is synonymous with actor and future filmmaker Orson Welles, who directed and narrated the H.G. Wells adaptation as an episode in the American radio drama anthology series *The Mercury Theatre on the Air*. It was performed as a Halloween episode of the anthology on October 30, 1938, and aired over the Columbia Broadcasting System network.

The first forty minutes or so of the hour-long show was broadcast as a run of simulated news bulletins. The first of these appeared to interrupt an airing of dance music to report that a set of curious explosions had been seen on Mars. This was soon followed by an apparently unrelated report of a strange object falling to Earth on a farm in Grover's Mill,

New Jersey. It transpires, of course, that Martians emerge from the fallen object to attack locals using a heat ray. What then followed was a rapid series of realistic reports of a dramatic and devastating alien invasion, happening live across the US and the rest of the world.

The realistic nature of the broadcast was helped by the fact that *The Mercury Theatre on the Air* was a radio show without commercial breaks. The first interlude in Welles's radio show came almost thirty minutes into the broadcast. In the days after Welles's adaptation, there was widespread media outrage about the broadcast. Welles's directorial use of realistic news bulletins was described as deceptive by some newspapers and commentators, and a call went out for the perpetrators of the broadcast to be disciplined by the regulating Federal Communications Commission. Nonetheless, the episode made Welles a famous dramatist.

But the adaptation also became famous for allegedly causing mass panic, which Kubrick referred to in his *Playboy* interview thirty years later. One theory is that many listeners tuned into *The War of the Worlds* during a musical break, missing the earlier announcement that the show was a drama.

You can't blame Kubrick. He was probably thinking of reports such as the 1940 book, *The Invasion from Mars*. In this study, Princeton professor Hadley Cantril figured that around six million people tuned into *The War of the Worlds* broadcast. Of this number, Cantril suggested that about 1.7 million listeners believed the broadcast was real and, of those, 1.2 million people were petrified or panicky.

Indeed, according to many common accounts, thousands of New Yorkers fled their homes from the Martian invasion. Swarms of petrified people were said to be flooding the streets in various US cities, trying to catch a glimpse of Martian Tripods and a real-life space encounter unfolding in front of their very eyes. Yet, in 1954, Ben Gross, radio editor for the *New York Daily News*, wrote in his memoir that the streets of New York were "nearly deserted" that night. The real extent of the panic seems to have come from a small group of Grover's Mill, New Jersey, residents. Thinking the town's water tower on Grover's Mill Road had been turned into a giant Martian war machine, as the radio show had said, residents fired guns filled with buckshot in a retaliatory attack on the water tower.

In 1998, one hundred years after Wells's original book, Grover's Mill residents held a light-hearted "Martian Ball" to commemorate the sixtieth anniversary of the incident.

DON'T PANIC!

In fact, not many Americans heard Welles's radio show live. Most people were tuned into the highly popular *Chase and Sanborn Hour*, a comedy show that was airing at the same time as *The War of the Worlds* on another radio station. Indeed, a radio ratings survey was, coincidentally, being carried out by telephone that night, a poll of approximately five thousand households. In response to the survey, it seems only 2 percent of people listened to *The War of the Worlds*. Not only that, but several major CBS affiliate radio stations weren't even broadcasting Welles's program, shrinking its potential audience still further.

There's little doubt Welles himself knew he was trying to instill a little panic over the Martians. He played recordings of Herbert Morrison's famous radio reports of the Hindenburg disaster to make more obvious the mood he wanted, but the main source of the alleged panic appears to have been fake news from the reporting newspapers. The *New York Times* ran with "Radio Listeners in Panic, Taking War Drama as Fact," the *Chicago Herald and Examiner* opted for "Radio Fake Scares Nation," and the *San Francisco Chronicle* declared "US Terrorized By Radio's Men From Mars." A regularly repeated claim was that within a month, 12,500 articles had been published worldwide on the alien mass panic, and yet in his detailed account of the reporting in his book, *Getting it Wrong*, American Professor W. Joseph Campbell found that most newspapers quickly ditched the story. (As the great Irish satirist Jonathan Swift wrote in 1710, "Falsehood flies, and the truth comes limping after it.")

It's far more likely the newspapers took the opportunity of Welles's broadcast to attack radio. The newspapers had a clear agenda. An editorial in the *New York Times* claimed that *The War of the Worlds* show was used in an attempt to censure the relatively new medium of radio—a fast-growing and serious rival in providing news and advertising. In *Slate* magazine's piece on the Martian mass panic in 2013, media professors Jefferson Pooley and Michael J. Socolow wrote, "How did the story of panicked listeners begin? Blame America's newspapers. Radio had siphoned off advertising revenue from print during the Depression, badly damaging the newspaper industry. So, the papers seized the opportunity presented by Welles's program, perhaps to discredit radio as a source of news. The newspaper industry sensationalized the panic to prove to advertisers, and regulators, that radio management was irresponsible and not to be trusted."

Given all the controversy, it's a bold move indeed to try copying Welles's 1938 broadcast, but that's exactly what Leonardo Paez and Eduardo Alcaraz did with their Spanish-language version for Radio Quito in Ecuador in February 1949. According to reports (it must be stressed!), the broadcast set off mass panic. Local police and fire services charged out of town to fight the hypothetical alien invasion. When it became clear this was yet another victory for the science-fictional imagination, the panic morphed into a riot, resulting in several deaths, including those of Paez's girlfriend and nephew. The offices of Radio Quito and El Comercio, a local newspaper that had contributed to the hoax by publishing reports of UFOs in the days before the broadcast, were both burned to the ground.

AVATAR:

ARE THERE OTHER EARTHS IN SPACE?

"Innumerable suns exist; innumerable earths revolve around these suns in a manner similar to the way the seven planets revolve around our sun. Living beings inhabit these worlds."
—Giordano Bruno, *On the Infinite Universe and Worlds* (1584)

"You are not in Kansas anymore. You are on Pandora, ladies and gentlemen. Respect that fact every second of every day. If there is a Hell, you might wanna go there for some R & R after a tour on Pandora. Out there beyond that fence every living thing that crawls, flies, or squats in the mud wants to kill you and eat your eyes for jujubes. We have an indigenous population of humanoids called the Na'vi. They're fond of arrows dipped in a neurotoxin that will stop your heart in one minute—and they have bones reinforced with naturally occurring carbon fiber. They are very hard to kill. As head of security, it is my job to keep you alive. I will not succeed. Not with all of you. If you wish to survive, you need to cultivate a strong, mental aptitude. You got to obey the rules: Pandora rules.
—James Cameron, *Avatar* screenplay (2009)

"The Earth is a very small stage in a vast cosmic arena. Think of the endless cruelties visited by the inhabitants of one corner of this pixel on the scarcely distinguishable inhabitants of some other corner, how frequent their misunderstandings, how eager they are to kill one another, how fervent their hatreds. Think of the rivers of blood spilled by all those generals and emperors so that, in glory and triumph, they could become the momentary masters of a fraction of a dot."
—Carl Sagan, *Pale Blue Dot:*
A Vision of the Human Future in Space (1994)

"We don't want to conquer the cosmos, we simply want to extend the boundaries of Earth to the frontiers of the cosmos. We don't want to enslave other races, we simply want to bequeath our values and take over their heritage in exchange. We are only seeking Man. We don't know what to do with other worlds. We are searching for an ideal image of our own world: we go in quest of a planet, of a civilization superior to our own but developed on the basis of a prototype of our primeval past.

—Stanislaw Lem, *Solaris* (1961)

THE ATOMISTS

The idea of other Earths is a notion that science fiction has kept alive for many centuries. The movie *Avatar* tells the tale of humans in the mid-twenty-second century colonizing the lush planetary moon of Pandora in the Alpha Centauri star system. The film is set two hundred years in the future, but let's take some journey two thousand years into our past. The trip will be worthwhile, as it will show just how long science and science fiction have influenced each other when it comes to the idea of other Earths.

The belief in other planets such as ours begins with the imaginative Atomists of the ancient Greek world. According to the Atomists, the universe was made of tiny countless uncuttable (*a-tomos*) particles. The Atomists also believed in the *void*: the empty space through which the atoms moved, without artifice. Injecting the notion of nothingness into philosophy was a daring move. Many early thinkers such as Aristotle considered the cosmos a plenum, a space entirely filled with invisible stuff or *horror vacui*; nature abhors a vacuum. But the Atomists believed that all you needed was atoms and void to explain nature's rich variety.

One of the notable Atomists was Democritus, a Greek thinker of the fifth century BC. Together with the founder of Greek Atomism, Leucippus, Democritus had a great influence on early cosmology and the study of the cosmos as a whole. Hippolytus of Rome, a third-century church scholar, recorded in his *Refutation of All Heresies* that the Atomists firmly believed that the cosmos contained many worlds:

"Democritus holds the same view as Leucippus about the elements, full and void . . . he spoke as if the things that are were in constant motion in the void; and that there are innumerable worlds, which differ in size. In some worlds there is no sun and moon, in others they are larger than in our world, and in others more numerous. The intervals between the worlds are unequal; in some parts there are more worlds, in others fewer; some are increasing, some at their height, some decreasing; in some parts they are arising, in others failing. There are some worlds devoid of living creatures or plants or any moisture."

The Atomists also had healthy democratic debates. This seems rather appropriate, as it was Democritus himself who gave the world the word democracy. For most Atomists, the belief in numerous other worlds also meant a universe replete with life. These other worlds would come and go, emerging and vanishing, some being born as others perished, in an eternal movement. This cosmic host of life-bearing worlds was beyond reach, however. By the term "world," the Atomists (as well as almost all other ancient authors) did not mean a different solar system, like our own, but one potentially visible from Earth. So other "worlds" were not exosystems: planets in orbit about distant stars. Rather, each was a self-contained cosmos, like our own, with an Earth at the center, and with planets and stellar vault surrounding. Maybe these "worlds" could be called other "realms," a lot like our contemporary "multiverse" idea, that speculative set of other universes, invisible and unreachable from our own universe. These other worlds of the Greeks could be contemporaneous with their ancient world or may form a linear succession in time.

The most famous Atomist was Epicurus. He championed the idea of a plurality of inhabited worlds in a number of writings, including a letter he wrote to one of his disciples, Herodotus. The letter illuminates Epicurus's views on other worlds:

"Furthermore, there are infinite worlds both like and unlike this world of ours. For the atoms being infinite in number . . . are borne far out into space. For those atoms, which are of such nature that a world could be created out of them or made by them, have not been used up either on one world or on a limited number of worlds, nor again on all worlds which are alike, or on those which are different from these. So that there nowhere exists an obstacle to the infinite number of worlds."

Atomism was more than just a cosmology. The idea of other Earths, and a universe full of life-bearing planets, was drawn from the very logic of their worldview. The Epicureans believed in an infinite number of worlds, built from an infinite number of atoms. The "world" of the Epicureans was "kosmos," a system of order, not of chaos. The kosmos was their observable universe, and they conjured up a stunning scheme of an infinite horde of such worlds, beyond the senses, but not beyond reason. The Atomist reasoning behind this cosmic picture was quite cultivated: as there must be an infinite number of atoms, and an infinite number of atoms could not have been exhausted by our finite world, so, as our world was born by chance collision of atoms in motion, other worlds must be forged in the same way.

Epicurus also believed these other Earths to be inhabited by life. In the same letter to Herodotus, he makes clear his early ideas of alien life:

"Furthermore, we must believe that in all worlds there are living creatures and plants and other things we see in this world; for indeed no one could prove that in a world of one kind there might or might not have been included the kinds of seeds from which living things and plants and all the rest of the things we see are composed, and that in a world of another kind they could not have been."

The godless worldview of Epicurus and the Atomists was not to everyone's liking. Take major thirteenth-century Italian poet, Dante Alighieri, for example. His great poem, *La Divina Commedia*, or *Divine Comedy*, was a journey through the Christian universe of Dante's days. The story of the *Divine Comedy* takes the reader through Hell (*Inferno*), Purgatory (*Purgatorio*), and Paradise (*Paradiso*). It is in Hell's inferno (Canto X,

Circle Six, "Where the heretics lie") where we find Epicurus and his Atomist followers. The Epicureans may have cherished friendship as a vital element of happiness—and their philosophical school may have been quite a cosmopolitan community by ancient standards, tolerating women and slaves and practicing vegetarianism—but they believed in a material model of the universe, a cosmos devoid of divine dabbling. Believing in other Earths, it seems, was a crime fit for damnation.

A TRUE STORY: GREEK ALIEN FICTION

The Atomists set the scene, with peerless logic and some ancient Greek gravitas. What better than a creative and imaginative fantasy to picture what kind of exotic life might dwell on these other worlds? Enter Lucian, an Assyrian orator and satirist who wrote in Greek. Lucian's ancient tale, *A True Story*, was a work of speculative fiction that just bristles with irreverent and wonderful ideas about what life might be like elsewhere in the cosmos. *A True Story* is the earliest-known fiction about other worlds and alien life. Like *Avatar*, Lucian's tale has themes of interplanetary imperialism and warfare but predates the likes of H.G. Wells and Jules Verne by seventeen centuries.

The story starts with a company of adventuring heroes taking a trip to the Moon. Lucian skillfully plays down the initial encounter with the Moon and other celestial bodies. The wealth of extraterrestrial wonders still to come, Lucian first focuses on our sense of familiarity, which serves to show that our world, from the narrator's point of view on the Moon, looks just like another land "below" and is otherwise identical to any other extraterrestrial planet.

On the Moon, the travelers find a war raging between the king of the Moon and the king of the Sun over the colonization of the Morning Star. As with *Avatar*, the armies that wage this war are staffed with the most exotic of recruits: cloud centaurs, stalk-and-mushroom men, and acorn-dogs or "dog-faced men fighting on winged acorns." Tellingly, the Moon, Sun, stars, and planets are depicted as realistic backdrops, each with its own distinct habitat and peoples. *A True Story* is a measure of significant changes in attitude and imagination due to Atomist ideas about other Earths in space.

A True Story is revolutionary. With this work, Lucian takes contemporary ideas and uses them to picture alternative worlds in a way that makes his readers look at things afresh. They become somewhat dislocated from their routine way of thought and are forced to contemplate other worlds and places. Like many who were to follow, Lucian seduces the reader into speculating on the habits of mind in the real world. Little wonder that Lucian's story is regarded by many as being the first true work of science fiction, the genre renowned for its depiction of alternative worlds radically unlike our own but somewhat similar in terms of scientific knowledge.

Like *Avatar*, Lucian's story vividly portrays other planets, especially his Moon world. He begins by describing the war between the Moon and the Sun as if they were expansive, mongrel empires of the ancient Greek world. Here we find the weapons of war, the considered tactics and strategies, and even scientific detail of a battle wound. Then, we get the anthropology and biology of his lunar world. The native people are described in terms of their clothing, their diet, and substantial detail of their physiology. In this way, the Moon world is presented as a parallel Earth, with all the same intricacies as the home planet. Like the fiction that was to follow in the wake of the scientific revolution centuries later, Lucian's text is about fictional worlds, but worlds that are developed within a scientific framework.

As an admirer of Epicurus, and like other Atomist philosophers, Lucian was a pluralist of the imagination. He had a questioning mind that traced back to the very oldest Greek thinkers, a curiosity about the wonderful variety in nature and human society alike. Lucian was a thorough skeptic. He was very aware that in many ways the old discarded myths were reemerging in the disguised form of some of the new philosophies, and the degeneration of thought into pseudoscience. Lucian's tale is the most ancient example of what has come to be known as the intellectual nonconformism of science fiction. The wealth of estranged worlds of *A True Story* was the shape of things to come.

BATTLESTAR GALACTICA AND STAR TREK:

WHY ARE WARS IN SPACE SO WRONG?

"The Beatles have no future in show business. We don't like your boys' sound. Groups are out. Four-piece groups with guitars, particularly, are finished."
—Decca Records executive to Beatles manager
Brian Epstein (1962)

"I see no good reason why the views given in this volume should shock the religious sensibilities of anyone."
—Charles Darwin, foreword to *On the Origin of Species* (1859)

"Next Christmas, the iPod will be dead, finished, gone, kaput."
—Alan Sugar (2005)

"Computers in the future may weigh no more than 1.5 tons."
—*Popular Mechanics* magazine (1949)

"A rocket will never be able to leave the Earth's atmosphere."
—the *New York Times* (1936)

"There is no likelihood man can ever tap the power of the atom."
—Robert Millikan, American physicist and Nobel Prize winner (1923)

BAD PREDICTIONS AND SPACE FLIGHT

Science fiction may have been right on the mark for predictions such as space travel, atom bombs, and mobile phones, but there were plenty of things on which the genre was embarrassingly off beam, with inventions such as schooling helmets, pills as meals, zombie invasions, and many more.

Take computers, for example. Old stories, written when computers were new and really quite huge, left writers struggling to imagine a future where the computer would become so small you could wear one on your wrist. As a result, tales often talked about *the* computer, as if each town or city only had just the one computer to be dealing with. It's a bit like the famous quote from the mayor of an American town who excitedly said about the newly invented telephone, "I can see a time when every city will have one."

Science fiction has been dreaming up Martians for well over a century, influencing the likes of American tycoon Percival Lowell, who thought in the 1890s that he saw Martian "canals" through the telescope. Early in the twentieth century when a French widow ran an essay competition about the question of life beyond the Earth, she barred the topic of Mars in her competition rules because it was "too obvious" that there was life on Mars. Essay writers were forced to pick another planet.

And, of course, there's the infamous jet pack. The logic of the jet pack went something like this: "Okay, planes are kind of like flying buses, yes? So why can't we each have a personal jet pack, which would be like a minihelicopter? We could call it the iJet!" Upon further reflection, perhaps, parents began to imagine their teenage kids with air rage, rather than road rage, bombing around uncontrollably in their iJets, as it would be practically impossible to police the iSky. Quietly they prayed the iJet would never happen.

Science fiction has also come up with some weird and wonderful ideas about space flight. As discussed elsewhere in this book, the spaceship was first imagined when Shakespeare was still alive, when Galileo first used the newly invented telescope as a weapon of discovery and Johannes Kepler at once dreamt up ships to sail the "aether." Kepler, who also wrote science fiction, used the power of imagination to conjure spaceships over 350 years before men landed on the Moon.

Now, it's one thing imagining yourself sailing away into space, but quite another dreaming up a way of actually getting there. Fiction writers had been doing this long before Kepler. Lucian, for example, had written in 160 AD about space voyages in his tale *A True Story*. Lucian's ingenious way of propelling his adventurers into space was by using a giant waterspout, which pushed them into the air so powerfully that they ended up in space. The trivial matter of how his travelers might actually breathe on their journeys in space doesn't seem to have worried Lucian.

In medieval times, perhaps a little concerned with the lack of steer on Lucian's waterspout idea, fiction writers considered methods a little less random. The Bishop of Llandaff, Francis Godwin, published his book *The Man in the Moone* in 1638. His preferred method of getting to the Moon was geese. Yes, Godwin's lunar voyager in the story harnesses himself to around three-dozen geese, which carry him higher and higher. He doesn't actually mean to end up on the Moon but remembers that the geese migrate each year to the moon to hibernate, and so it is to the Moon they carry him and where his adventure then unfolds.

If geese are not to your liking, there's always dew. Maybe the most inventive method of propulsion was dreamt up by romantic French hero Cyrano de Bergerac. His reasoning went something like this: dew droplets can be seen first thing in the morning, but when sunlight shines on them, they evaporate, rising into the air, so the hero in Cyrano's story collects as many dew droplets as he can, bottles them, and securely fastens the bottles around his waist. When he then stands in the morning sunlight, hey, presto, the evaporating dew lifts him to the Moon!

SPACE WARS

H. G. Wells had the Martians carry war to the Earth in *The War of the Worlds*, one of the most influential of all early science-fiction stories, but the power that science fiction has on our idea of war in space is so strong that it's hard to imagine the actual reality of what war might look like in the military theater of the heavens above. You've seen the kind of war in space that science fiction dreams up. In *Battlestar Gallactica* or *Star Trek*, we get pretty pictures of colored lasers or photon torpedoes pinging their way through the black as sleek spaceships scream through empty space. As fact has often had to take a backseat to the fiction in film and television, a realistic war in space has seldom been seen on-screen. Instead, the huge sway of science fiction over the last half century of sci-fi culture has scorched into our subconscious an expectation of swift-moving and winged spacecraft, luminous laser-fire, and instantaneous communication between space warriors.

A real space war would look very different indeed. Consider the winged spacecraft. From the X-Wing in *Star Wars* to the Klingon Battle Cruisers in *Star Trek*, many a fictional ship sports go-faster flaps, yet a spaceship has no need for wings save the odd excursion down to its home planet in a thicker atmosphere. Wings are only required when a ship needs lift in the sock of air that surrounds a planet. In reality, a spaceship could take any shape. Clarke and Kubrick realized this when they made *2001: A Space Odyssey*. The movie's ship, Discovery One, was an interplanetary spaceship based on solidly considered yet unrealized science consisting of a 52-foot habitat sphere at its head and a long spine of storage tanks. Discovery One was based on a set of NASA-funded studies done in 1962 called EMPIRE (Early Manned Planetary-Interplanetary Round-Trip Expedition), which produced a number of concept vehicles. Kubrick made sure that an early production prototype of his fictional Discovery, which proposed "wings" to radiate heat, was rejected, as moviegoers might take the wings to mean that Discovery was meant to fly through an atmosphere.

So, wings are out. What about the question of intership or inter-planetary communication? In most science-fiction movies, this seems instantaneous. And yet, imagine a space war breaks out somewhere in the vicinity of Mars. Don't say you're surprised. Battles always seem to be breaking out in the Martian neighborhood. The planet Mars, after all, is named after the God of war, and its moons are named after the characters Phobos (panic/fear) and Deimos (terror/dread), who, in Greek mythology, went with their warring father into battle. Consider Mars as one of Earth's colonies under alien attack. If mission control back on Earth is trying to communicate with our Martian Armada, in reality there's going to be a time delay, and that delay very much depends on how far away Mars is at the moment. If Mars is on the same side of the Sun as Earth (quick Copernican update: both Earth and Mars orbit the sun), then the communication delay will be around four minutes, but if Mars is on the far side of the Sun from Earth, the delay will be a maximum of around twenty-four minutes. In short, talking to off-ship space warriors during a battle would be complicated and slow! It's for this reason that the great American science-fiction author Ursula K. Le Guin invented the ansible. This fictional gadget's name was a contraction of the word "answerable," and the device allowed users and space warriors alike to talk instantaneously over interstellar distances.

"In space no one can hear you scream" is one of the most quoted and parodied taglines in all of film history, not just science fiction, and yet science-fiction moviemakers don't seem to have learned the lesson of this tagline from Ridley Scott's seminal 1979 movie, *Alien*. All too often we hear space explosions in the movies. From Death Star to droid, they often go *boom*. But, as there is no air in space to carry the pressure waves that make up sound, the "sound" of destruction would actually be silence.

Then, there's the actual fighting. All ships have mass, and all masses have a quality known as momentum: mass times velocity. The momentum of a body is conserved, so if you fire mass away from you in the form of weaponry, it will either speed you up or slow you down. In science-fiction films, spaceships mostly fire forward in the same direction that they're moving, but if the ships are also trying to accelerate, the firing would result in a loss of velocity because of the conservation of momentum. In a real war in space, ships would be more likely to fire in other directions, as losing velocity would mean wasting precious fuel. Real space wars would involve weird-shaped ships, firing off projectiles silently in vectors that wouldn't lose them velocity and picking up mission commands either from a newly invented ansible or on a time delay according to some preagreed strategy from mission command.

One final way in which science fiction has messed with our minds? In space, there's no right way up, and no down, so ships wouldn't simply line up in space in some kind of cosmic face-off. A real space war would be far more mind-bogglingly random!

THE MARTIAN:

WHEN WILL WE COLONIZE SPACE?

"We are all tired of being stuck on this cosmical speck with its monotonous ocean, leaden sky, and single moon that is half useless. Its possibilities are exhausted, and just as Greece became too small for the civilization of the Greeks, so it seems to me that the future glory of the human race lies in the exploration of at least the solar system!"

—John Jacob Astor, *Journey in Other Worlds: A Romance of the Future Story* (1894)

"A manuscript I wrote . . . speculated as to the last migration of the human race, as consisting of a number of expeditions sent out into the regions of thickly distributed stars, taking in a condensed form all the knowledge of the race, using either atomic energy or hydrogen, oxygen and solar energy . . . [It] was contained in an inner envelope, which suggested that the writing inside should be read only by an optimist."

—Robert Goddard, *Material for an Autobiography* (1927)

"There is no way back into the past; the choice, as Wells once said, is the Universe or nothing. Though men and civilizations may yearn for rest, for the dream of the lotus-eaters, that is a desire that merges imperceptibly into death. The challenge of the great spaces between the worlds is a stupendous one; but if we fail to meet it, the story of our race will be drawing to its close."

—Arthur C. Clarke, *Interplanetary Flight* (1950)

"The dinosaurs became extinct because they didn't have a space program. And if we become extinct because we don't have a space program, it'll serve us right!"

—Larry Niven, as quoted by Arthur C. Clarke
in "Meeting of the Minds: Buzz Aldrin Visits
Arthur C. Clarke" by Andrew Chaikin (2001)

"My father, Edwin Eugene Aldrin, was an engineer and an aviation pioneer—and a friend of Charles Lindbergh and Orville Wright. . . . I have a message in a time bottle for the candidate who wins the 2016 election for the US presidency. There's opportunity to make a bold statement on the occasion of the July 2019 50th anniversary of the first humans to land on the moon: 'I believe this nation should commit itself, within two decades, to commencing American permanence on the planet Mars.'"

—Buzz Aldrin, *Mission to Mars:
My Vision for Space Exploration* (2013)

SHIPS INTO SPACE

Mark Watney's journey in *The Martian* movie began with ships. Starting with the medieval voyages of discovery, the ship had become a kind of weapon of exploration. German writer Bertolt Brecht knew all about the power of the ship when he wrote his famous play about the *Life of Galileo*. In the very first scene, Brecht put these words into the mouth of the renowned Tuscan astronomer:

"For two thousand years men have believed that the Sun and the stars of heaven revolve around them. . . . The cities are narrow and so are men's minds. Superstition and plague. But now we say: because it is so, it will not remain so. I like to think that it all began with ships. Ever since men could remember they crept only along the coasts; then suddenly they left the coasts and sped straight out across the seas. On our old continent a rumor started: there are new continents! And since our ships have been sailing to them the word has gone 'round all the laughing continents that the vast, dreaded ocean is just a little pond. And a great desire has arisen to fathom the causes of all things. . . . For where belief has prevailed for a thousand years, doubt now prevails. All the world says: yes, that's written in books but now let us see for ourselves."

So began the great days of empire when the agenda for science, exploration, and trade became increasingly blurred. Science and the ship were of prime importance. To build an empire, you needed to know *where* you were and *what* you owned. Since all the best trade routes were wet, commercial control very much depended on the speed and reliability with which long-range voyages took place along those trade routes. The discovery of the Americas and the quest for longitude and dominion in that new world promoted a greater need for science. And one of science's first weapons of discovery was the ship itself.

Galileo's telescope became a kind of ship. Just think about the way in which he used it. Galileo took himself as observer, and his contemporary world as onlookers, to a place only science-fiction writers had ever imagined. But, like the destinations of a ship, the destinations of the telescope also became public knowledge. If you didn't believe what Galileo said, you could take a look for yourself. If the ship had gifted evidence of new worlds on Earth, surely the telescope would soon present evidence of new worlds in space?

THE OUTWARD URGE

The first science-fiction stories of the medieval age were all about voyages into space. It began with Johannes Kepler and *Somnium*, his imagined journey to the alien world of the Moon. Kepler didn't limit his imagination to fiction, but he was hopeful that space wasn't populated by the same pirates that plagued the seven seas on Earth! He celebrated the fact that "navigation across the vast ocean is less dangerous and quieter than in the narrow, threatening gulfs of the Adriatic, or the Baltic, or the British straits" and urged the world to "create vessels and sails adjusted to the heavenly ether."

And so, in a very real sense, the idea of the human colonization of space is over four hundred years old. Nineteenth-century speculations about colonizing the other worlds of our solar system had an uncertain history. This comes down to two facts. First, discoveries in science and astronomy had the habit of contradicting the optimism of the science-fiction imagination. Hard facts got in the way of blue-sky thinking. Second, a good number of the early science-fiction writers were British, and the example of the British Empire was enough to inspire cautionary tales about the human extension into space, British or otherwise. Under such clouds, H.G. Wells considered the example of Britain's colonial history as an analogy for the Martians' conduct in *The War of the Worlds*. Wells deliberately engaged in the sense of shame about the methods employed in colonizing Earth. In particular, Wells drew a strong parallel between the fictional Martian invasion of Earth and the factual European invasion of Tasmania. The sheer brutality of the politics of colonization has been a key issue in science fiction ever since.

Wells didn't seem to consider the concept of the human colonization of Mars. Other writers, for similar reasons, imagined humans colonizing other worlds only under duress, as Earth became uninhabitable through cometary impact, population explosion, or the searing heat of an expanding Sun. A science-fictional obsession with Mars began for those who dared dream because it's the closest planet to Earth whose surface can be easily sighted through our scopes. Mars was always a major target for its polar caps, seasons, and a twenty-four-hour day. It inspired new modes of space living, which needed to be imagined, researched, and developed.

Some considered the Moon to be the first step. In his 1951 story, *The Man Who Sold the Moon*, American writer Robert Heinlein imagined a venture capital mission to the Moon. In his tale, once the first flights had been successful, a Moon colony was next. You can see the lunar logic. The Moon is on our cosmic doorstep. It's within easy reach of Earth, enabling pioneers to exchange goods and services with our home planet, and it's also a stepping-stone for further exploration to Mars and beyond, as well as being a base for telescopic research.

Heinlein predicted a lunar landing in 1978. The Apollo program got there a decade early. Their missions enabled human life on the Moon and tested the challenges of lunar living. They discovered huge temperature swings, from 250°F at noon to -250°F at night, and a surface constantly blasted by micrometeorites and cosmic rays. Each mission was longer than the last to check human endurance. It became clear that, to survive such bombardment, colonists would have to live underground in "lava tubes," long, natural caves formed by flowing magma in the past.

Apollo lunar science experiments taught us much about future exploration, such as how to transport large items of equipment and materials. One plan is to put a farm at the Moon's north pole, where there's eight hours of sunlight each day, throughout the summer. The farm would need radiation protection and insects for pollination, but a farm measuring only half a mile could feed 100 people.

MAKE LIKE MARK WATNEY

What we learn on the Moon can be adapted for Mars, as we saw with Mark Watney's experience in *The Martian* in which a colony on the Red Planet fends off global dust storms, big swings in temperature, and harmful solar radiation. An even bigger task would be to melt the polar ice into a sea 12 meters thick, covering much of the planet.

How do scientists and engineers intend to go about colonizing Mars?

It's a commonly held belief that the first launch could carry an unmanned Earth Return Vehicle (ERV) to Mars. The ERV would contain a nuclear reactor, which would power a unit to make fuel using material found in the Martian atmosphere. Two years later, a manned mission would touch down near the ERV. The crew would stay for eighteen months, exploring the planet until returning to Earth using ERV-made fuel. The crew would be replaced by another team, and a string of bases would be set up.

At first, pioneers would depend on goods from Earth. The Martian biodomes would let in sunlight and screen out the harmful solar rays. Space suits would not be needed, as pioneers could roam in oxygen masks and protective clothing. This way, pioneers could leave their domes and build other domed villages and farms.

Mars' soil must be transformed, or terraformed. Special units would pump gases such as methane and ammonia into the atmosphere. The gases would absorb solar energy and warm the planet, triggering the release of carbon dioxide from the soil and ice caps. The carbon dioxide in the atmosphere would help global warming, so oceans would form. After several decades of terraforming, Mars might look as blue and watery as Earth. Within a century, it could be terraformed into an oxygen-rich environment, supporting a human colony, some of whom may dream of traveling to the remote corners of the solar system and beyond.

PASSENGERS:

WILL SPACE TOURISM EVER PLEASE JENNIFER LAWRENCE?

"Space is big. Really big. You just won't believe how vastly, hugely, mind-bogglingly big it is. I mean, you may think it's a long way down the road to the chemist, but that's just peanuts to space."
—Douglas Adams, *The Hitchhiker's Guide to the Galaxy* (1978)

"Well, now you've put your finger right on it. In order to have all of these wonderful things in space, we don't have to wait for technology—we've got the technology, and we don't have to wait for the know-how—we've got that too. All we need is the political go-ahead and the economic willingness to spend the money that is necessary. It is a little frustrating to think that if people concentrate on how much it is going to cost they will realize the great amount of profit they will get for their investment. Although they are reluctant to spend a few billions of dollars to get back an infinite quantity of money, the world doesn't mind spending $400 billion every year on arms and armaments, never getting anything back from it except a chance to commit suicide.
—Isaac Asimov, interview with Phil Konstantin,
Southwest Airlines magazine (1979)

"Since, in the long run, every planetary society will be endangered by impacts from space, every surviving civilization is obliged to become spacefaring—not because of exploratory or romantic zeal, but for the most practical reason imaginable: staying alive."
—Carl Sagan, *Pale Blue Dot:*
A Vision of the Human Future in Space (1994)

"If Earth is considered a closed system, there will be less for all forever. The frontier is closed, the wilderness is gone, nature is being destroyed by human consumers, while billions are starving. The future indeed looks grim, and there are, ultimately, no really long-range, positive solutions, nor motivation for making the sacrifices and doing the hard work needed now, unless we understand that we are evolving from an Earth-only toward an Earth-space or universal species."

—Barbara Marx Hubbard, *Distant Star* (1997)

I am a passenger
I stay under glass
I look through my window so bright
I see the stars come out tonight
I see the bright and hollow sky . . .
The sky was made for us tonight

—Iggy Pop, "The Passenger" (1977)

SPACE IS BIG

As Douglas Adams once said in *The Hitchhiker's Guide to the Galaxy*, space is big. Sure, it starts really close to home. Space begins only 62 miles away. In the grand scheme of things, that isn't very far at all. If you live in Seattle, Canberra, Hyderabad, Cairo, Beijing, or central Japan, for example, space would be closer to you than the sea. When trying to get your head around the sheer size of space, and the huge distances we need to cross in the future, it helps to remember that moons go around planets, planets go around stars, stars are arranged into huge galaxies, and galaxies make up the large-scale structure of the universe.

The size of the cosmos is unimaginably huge. That's why astronomers use the speed of light to measure the universe. You know the drill: as light is the fastest thing known to us, it's a useful gauge of huge distances. Light takes a second to reach the moon, eight minutes (roughly speaking) to reach the sun, and four years to get to the next closest star. A beam of light travels about 186,000 miles in one second. In the Monty Python "Galaxy" song, they talk about light traveling at "twelve million miles a minute," but this is surely with some artistic license, as the actual figure is 11,176,943.82. Light can travel about 6 trillion miles in a year. Our Milky Way Galaxy is so large that even light takes 100,000 years to cross it. And a recent estimate of the number of galaxies in the universe is at least two trillion. You can see why Douglas Adams was happy to whittle down his explanation to three words: space is big.

The trouble is, humans don't travel at light speed. The makers of the movie *Passengers* were very aware of this fact. They weren't overly ambitious about their sojourn into space. In the film, the starship Avalon transports colonists and crew in hibernation pods to the planet Homestead II, a journey taking only 120 years.

As talked about elsewhere in this book, science fiction has long dreamed of venturing into space. So far, however, only a few hundred astronauts and cosmonauts have made the trip. That may soon change. Companies are currently planning space tourism projects, as long as you have the money to pay for the flight. In the movie *Passengers*, too, private companies have ventured into the exploration of space. An enterprise called the Homestead Company finds, grooms, and eventually populates exoplanets as colonies. That's some pretty impressive real estate work, as the company would have to do research at an advanced level to locate exoplanets in orbit about relatively near stars.

SPACE TOURISM

One of the fictional front-runners for the idea of space tourism was British science-fiction writer Arthur C. Clarke. He imagined businesses working on the Moon in *A Fall of Moondust* in 1961 and partnered with American film director Stanley Kubrick on the famous 1968 movie, *2001: A Space Odyssey.* The movie was one of the very first to carry "product placements," featuring companies such as Bell, IBM, Pan Am, and AT&T. It was a prophetic vision of future space travel, complete with corporate logos and trademarks, showing a world absolutely managed by private capital. *A Space Odyssey* also dreamed up space tourists being served drinks by a stewardess in a station orbiting the Earth. It all looked so very civilized, at least on the surface.

Back in reality, the possibility of space tourism got a lot closer in 2004 with the Ansari X-Prize. This was a competition to design a reusable way of taking people into space, getting them back to Earth, and relaunching again within two weeks. The prize was won, the technology is within reach, and now companies are developing flights into space for the twenty-first century ahead. At the moment, of course, there's hardly anywhere to actually go, at least not until a tourist space station is built such as the one in *A Space Odyssey*. For now, passengers will be taken just above the 62-mile barrier that separates Earth from space, and then they will be brought back home. It's much like a helicopter trip around the Statue of Liberty, only darker and more expensive.

Science fiction first started dreaming about space stations in the nineteenth century. As humans have been used to the familiar sight of the Moon gracing our night sky, writers began to imagine adding an artificial satellite to Earth's natural satellite. An early artificial satellite of this kind was referred to as a "brick moon" in a story of that name written by Edward Everett Hale in 1861. French writer Jules Verne was also imagining satellites in his 1878 story, *The Begum's Millions*. In this book, a projectile is shot from a huge cannon with so much force that it enters Earth's orbit. One of the characters in the story writes in a letter, ". . . we saw your perfect shell, at forty-five minutes and four seconds past eleven, pass above our town. It was flying towards the west, circulating in space, which it will continue to do until the end of time."

Companies are now actively dreaming up space stations to place in orbit about the Earth. Bigelow Aerospace is working on the idea of an inflatable space hotel. Imagine taking a dip in a space station swimming pool with a glass bottom through which you can see the Earth below. Let's hope gravity loss doesn't cause the kind of giant-blob-of-water out-of-pool experience that happens to Jennifer Lawrence in *Passengers*. Orbital stations can be made to spin, which makes a kind of gravity that grounds those inside. Such spins can be spotted in *Passengers* on the Avalon, as well as the Cloverfield Station in *The Cloverfield Paradox*. Stations would get the same sunlight as the Earth and can be built with dome sections where food could be grown. These space cities could house hundreds or even thousands of people in the future. Companies are no doubt waging that such space experiences will feel more like a proper holiday in space, though perhaps not enough to please Jennifer Lawrence.

At a press day for the *Passengers* movie, screenplay writer Jon Spaihts sat on a science panel. He spoke about the technologies currently being developed to make long-distance space travel a reality and talked about some of the fictional tech featured on Avalon in the movie: "There is no warp drive, hyperspace or artificial gravity. It runs on fusion. It is propelled by a constant thrust ion drive and probably gets up to high speed with the aid of some booster or launcher that lobs it out to get it moving. But then, after that, it's a fractional G constant thrust ion drive. It has a kind of meteor screen at the front which is probably electromagnetic, but I imagine it has some aspects of a Bussard ramjet, meaning it is collecting space gas as it flies for fusant and for ion propellant. It's actually harvesting mass and it solves the propellant problem in that way by harvesting the trace quantities of space dust and gas that exist everywhere." The ramjet is another invention from the history of science fiction. According to Arthur C. Clarke's book, *Greetings, Carbon-Based Bipeds*, Cyrano de Bergerac invented the ramjet in 1657 in his novel *The States and Empires of the Moon*, where Cyrano wrote the following:

> "I foresaw very well, that the vacuity . . . would, to fill up the space, attract a great abundance of air, whereby my box would be carried up; and that proportionable as I mounted, the rushing wind that should force it through the hole, could not rise to the roof, but that furiously penetrating the machine, it must needs force it upon high."

Finally, consider the space elevator. If space tourism and space stations are to become a reality, we'll need a cheap and regular way of getting back and forth into space, right? Some way that doesn't use up too much rocket fuel and is kind to the environment. That's where the space elevator comes in. Imagine jumping into an elevator and pressing the button marked "space" or even "space station." Such a thing was imagined in an 1895 book called *Daydreams of Heaven and Earth*, by Russian rocket pioneer Konstantin Tsiolkovsky, who wrote science fiction as well as science.

At heart, the design of a space elevator is a 30,000-mile-long cable made from an as yet undeveloped material. This "beanstalk" material would need to be thirty times stronger than steel and have a diameter of no more than ten centimeters. Tricky. This superstrength beanstalk stuff would be tethered to the space station to enable continuous transport into space. NASA has spent millions of dollars running a competition to design space elevators. To date, the space elevator idea has failed to capture the public imagination. And yet the world, and Jennifer Lawrence, needs a space elevator. Our future in space demands the cosmic power behind an elevator button marked "space."

PART II
TIME

The science fiction of time has had a huge influence on us. Modern culture is obsessed with time. We wonder what might happen if we could tamper with time, or if we could jump time lines, guide evolution, rewrite history, or even cheat death. Science fiction is all about the relationship between us humans and the nonhuman but natural world around us, as revealed by science. Ever since the scientific revolution around five hundred years ago, science has encroached upon all aspects of life. Science has sought not just to explore, but to exploit nature. To master it.

Around the seventeenth century, we began to realize that time was limitless and inhumanly vast in scale. By the time the Industrial Revolution was well under way, great machines were turning over the soil of the world. Dinosaurs were discovered, and the death toll of extinction rang out for the first time. The fossil record churned out evidence of creatures no longer found on Earth. The new theory of evolution forced us all to confront the terrible extent of history. What if humans too became extinct? Little wonder *Mad Max* portrays a species running down into chaos.

Suddenly, there was no greater challenge for science. What if we could master time, the brutal agent that devours beauty and life? And so, our science-fictional obsession with time began. True, there had been folkloric flirtations with time: time-slip romances where dreamy magic is mixed with myth and time is lost as a convenient plot device. The idea of mechanized time travel, however, did not appear until industrialization. Its invention was tied up with the concept of time itself. The ancient

Greeks had two words for time, *kairos* and *chronos*. *Kairos* suggested a moment of time in which something special happens. *Chronos* was more concerned with measured, sequential time. Industrial society brought a mechanistic approach to nature. *Chronos* came to the fore, and time travel was born.

H.G. Wells gave science fiction one of its most enduring devices: the time machine. The long road to *Looper* and a thousand other fictional time machines all started with Wells and his novella, *The Time Machine*, an ingenious voyage of discovery, in which the time traveler sets out to marshal and master time, but he discovers the inevitable truth; time is lord of all. The significance of the story's title becomes clear. Humans are trapped by the mechanism of time and bound by a history that leads to inevitable extinction. Even the very stars themselves grow old and die.

Nonetheless, science-fiction writers still continued in their quest to master time. To nail down the future for us. Wells himself wondered what the future held for humans in *The War of the Worlds*. The invading Martians are not only a brutal force of evolution, they are also the "men" of the future. They are alien, yet they are human. They are what we may one day become, with their overdeveloped brains and emaciated bodies. They are the tyranny of intellect alone.

Evolution, of course, is a process that reveals itself in time. In his all-time classic movie, *2001: A Space Odyssey*, Stanley Kubrick suggested that the dumb, blind evolution of man would have to be interrupted by the guiding hand of an alien civilization to rescue us from the long, pathetic road to racial extinction.

Time travel proved to have further potential. In film and fiction, it soon became common for characters to jump freely between alternative time lines, each time line associated with its own plausible future. Such is the case with the *Terminator* series, in which apparently inescapable cyborg assassins are sent back from the future by a race of artificially intelligent machines bent on the extermination of humans!

Inevitably, the main focus of such time-traveling science fiction is melodrama, and yet science fiction has greatly influenced our culture, inspiring questions such as: How open is the future? Do we really have free will? Isn't all history in a sense a fiction? How can we ever know anything about time other than the fables we create? After all, we are all stories in the end.

BACK TO THE FUTURE:
WILL TIME TRAVEL EVER BE POSSIBLE?

"If time travel is possible, where are the tourists from the future?"
—Stephen Hawking, *A Brief History of Time* (1988)

"If we could travel into the past, it's mind-boggling what would be possible. For one thing, history would become an experimental science, which it certainly isn't today. The possible insights into our own past and nature and origins would be dazzling. For another, we would be facing the deep paradoxes of interfering with the scheme of causality that has led to our own time and ourselves. I have no idea whether it's possible, but it's certainly worth exploring."
—Carl Sagan, interview, *NOVA*, 12 October, 1999

"How do you decide whether tomorrow's technology includes time travel? Where do you look for evidence that our descendants have discovered the means of temporal voyaging? If time travel is a one-way process forward, there is no way we can know. If, as the new physics suggests, it is possible to move back in time, then the evidence we are searching for will present itself as anachronisms. Human beings are careless. They drop things they shouldn't, like the metal tubes found in Saint-Jean de Livet in France. They are also vulnerable. Whatever safeguards are in place, sooner or later someone will be trapped in a time period other than their own and die there. If the time period is historical, their death will leave no anachronistic trace, but if we examine the depths of prehistory, it becomes possible to trace the series of temporal disasters, which left a trail of corpses where they decidedly should not be.
—J.H. Brennan, interview, *Time Travel: A New Perspective* (1997)

"There have been mountains of nonsense written about traveling in time, just as previously there were about astronautics—you know, how some scientist, with the backing of a wealthy business-man, goes off in a corner and slaps together a rocket, which the two of them—and in the company of their lady friends, yet—then take to the far end of the Galaxy. Chronomotion, no less than Astronautics, is a colossal enterprise, requiring tremendous investments, expenditures, planning."

—Stanislaw Lem, interview, *The Star Diaries* (1957)

"The time when Christ was 33, so I could see if he really got up from the grave. Or maybe I'd go back to the time of Cleopatra, before Marc Antony got tight with her. I'd try to hit on her first.

—Muhammad Ali, on when in history he'd visit,
"The Ali Mystique," *Newsweek* (1975)

WRINKLES IN TIME

Imagine Muhammad Ali standing at the Crucifixion, amazed, open-mouthed, and staring at maybe the most famous scene in all of history. It was an expensive package, but the time travel tour guide said it was the only way Ali could check whether Christ really did get up from the grave. If this goes well, Ali will soon move swiftly on to his next trip—Cleopatra's Alexandria. But, for now, just one main point to remember: Ali must do nothing to disrupt history. When the call went out, asking who in Roman custody should be freed, Ali did as bid and joined with the crowd to holler, "Give us Barabbas!" It was then that Ali realized something strange about the crowd. Not a single soul from 33 AD was present. The mob condemning Christ to the cross is made up lock, stock, and smoking barrel of time tourists from the future.

This nifty time-travel scenario (not including the Ali twist) was dreamed up in the 1975 science-fiction story *Let's Go to Golgotha* by Garry Kilworth. It's typical of the kind of conundrum that writers have to reckon with when they tell a tale about tampering with time. The time tourists from the future litter the entire scene at Golgotha. They change the outcome of history itself, by not only being present at the Crucifixion, but also by influencing the outcome. The time tourists think they know the way the story is meant to go—they're meant to free Barabbas, the insurrectionary bandit—but the decision only goes that way because the time tourists are witness to the scene. Would Jesus have been set free instead, if they hadn't interfered? Such are the paradoxes when you tamper with time. It's why Carl Sagan suggested time travel would make history an experimental science.

This type of whacky paradox, which is often thrown up in time travel, is no doubt one of the reasons Professor Stephen Hawking refused to believe time travel was possible. His argument goes something like this: "If time travel really IS possible, then where are the time tourists of the future? Why aren't they visiting us, telling us all about the joys of time travel?' This Hawking scenario was taken a little further by Brian Clegg in his 2011 book, *How to Build a Time Machine: The Real Science of Time Travel*, in which Clegg asks, "If time travel is possible, why not flag a certain place and time in history and invite time travelers to attend? As long as information on the event percolated into the future—and a combination of Internet, print media, and TV coverage would seem to guarantee this unless our civilization were destroyed—how could any time traveler resist?"

British physicist Paul Davies has a solution. And it involves wormholes. As Davies puts it in his own *How to Build a Time Machine* of 2001, "A much-voiced objection to travel backwards in time is that we don't encounter . . . key historical events such as the Crucifixion . . . crowded by throngs of eager witnesses. . . . Fortunately this objection is easily met in the case of wormhole time machines. Although wormholes could be used to go back and forth in time, it is not possible to use one to visit a time before the wormhole was constructed. If we built one now, and established, say, a one-hundred-year time difference between the two ends, then in one hundred years someone could revisit 2001. But you couldn't use the wormhole to go back and see the dinosaurs. Only if wormhole time machines already exist in nature—or were made long ago by an alien civilization—could we visit epochs before the present. So, if the first wormhole time machine were built in the year 3000, there could not be any time tourists in the year 2000."

WORMHOLES AND WARPED SPACE

What's all this about wormholes? Time travel must surely be a long-held human aspiration, at least in spirit. And plenty of tales proliferate in which time travel is a major part of the plot, such as the temporal festive hijinks in Charles Dickens's *A Christmas Carol* (1843), and the burlesque and ludicrous book that is Mark Twain's *A Connecticut Yankee in King Arthur's Court* (1889). However, such stories and adventures usually begin with dreams, magic, or a blow to the head before the time travel unfolds.

Modern notions of time travel, with all its complexities, date back to 1905, and famous Nobel Prize-winning scientist Albert Einstein's special theory of relativity. His special theory showed that time and space are closely linked, and his general theory of 1916 showed that space and time are pliable. In Einstein's universe, spacetime can be warped, bent, expanded, or contracted in the presence of energy or matter. That means if you fill space with an exotic form of energy, it can be warped so that space and time bend back on themselves, allowing a potential traveler with the right kind of tech to tinker with time.

Science fiction got there first. Famous American writer John Campbell was the man who invented "space warps." In his 1931 story, *Islands of Space*, Campbell used the idea as a shortcut from one region of space to another. He followed that up with his 1934 story, *The Mightiest Machine,* in which he called this same shortcut "hyperspace," another now-familiar phrase. It was a year later that Einstein, with his colleague Nathan Rosen, came up with the science fact behind the science fiction of such time travel and worked out the scientific theory that explained the notion of "bridges" in space. It was much later that scientists started calling these bridges "wormholes."

What would such a wormhole look like? If you've ever watched *Doctor Who,* then you've probably seen the swirly cosmic tunnel down which the TARDIS often disappears when on another of its journeys through time and space. A wormhole has at least two mouths connected to a single throat. Stuff may "travel" from one mouth to the other by passing through the wormhole. We haven't found one yet, but the universe is immense, and we haven't really been looking for very long.

Scientists really do believe wormholes exist, at least in theory. As that theory belongs to Einstein, it takes a brave scientist to disagree. However, American cosmologist, Lawrence Krauss, author of the 1995 book *The Physics of Star Trek,* suggested on NBC's *What Einstein and Bill Gates Teach Us About Time Travel* in May 2017, "Most physicists now working would bet against the possibility of time travel, not merely because of the practical difficulties of generating the necessary conditions to allow it, but also because of the implications of time travel if it becomes possible."

To date, the idea of wormholes in our collective imagination owes almost everything to science fiction. However, if you fancy your chances of making that fiction fact, here's a short wormhole recipe for you. Take a dash of exotic matter. Make sure this matter is made up of particles that have antigravity properties. Pop them into the throat of a wormhole. (Note: beware black holes, which are one-way journeys to oblivion.) Your wormhole should have two mouths, an exit, and an entrance. Your challenge will be to keep the wormhole's throat open by using a force opposed to gravity, kind of antigravity, if you like. Boom, the wormhole throat stops imploding, and you've made a time machine! The stuff of science fiction has once again become fact.

INTERSTELLAR:

HOW DOES TIME WORK AS A DIMENSION?

"Time exists not by itself; but simply from the things which happen, the sense apprehends what has been done in time past, as well as what is present, and what is to follow after."
— Lucretius (c. 70 BC), as quoted by
William Hurrell Mallock, *Lucretius* (1883)

"Absolute, true and mathematical time, of itself, and from its own nature, flows equably without regard to anything external, and by another name is called duration: relative, apparent and common time is some sensible and external (whether accurate or unequable) measure of duration by means of motion, which is commonly used instead of true time; such as an hour, a day, a month or a year."
— Isaac Newton, *Principia* (1687), as quoted in Andrew Motte's.
The Mathematical Principals of Natural Philosophy (1803)

"Minkowski's idea and the solution of the twin paradox can best be explained by means of an analogy between space and spacetime . . . Time as a fourth dimension rests vertically on the other three—just as in space the vertical juts out of the two-dimensional plane as a third dimension. Distances through spacetime comprise four dimensions, just as space has three. The more you go in one direction, the less is left for the others. When a rigid body is at rest and does not move in any of the three dimensions, all of its motion takes place on the time axis. It simply grows older . . . The faster he moves away from his frame of reference . . . and covers more distance in the three dimensions of space, the less of his motion through spacetime as a whole is left over for the dimension of time. . . . Whatever goes into space is deducted from time. . . . In comparison with the distances light travels, all distances in the dimensions of space, even those involving airplane travel, are so very small that we essentially move only along the time axis, and we age continually. Only if we are able to move away from our frame of reference very quickly, like the traveling twin . . . would the elapsed time shrink to near zero, as it approached the speed of light. Light itself . . . covers its entire distance through spacetime only in the three dimensions of space. . . . Nothing remains for the additional dimension . . . the dimension of time. . . . Because light particles do not move in time, but with time, it can be said that they do not age. For them "now" means the same thing as "forever." They always "live" in the moment. Since for all practical purposes we do not move in the dimensions of space, but are at rest in space, we move only along the time axis. This is precisely the reason we feel the passage of time. Time virtually attaches to us.

—Jürgen Neffe, *Einstein: A Biography* (1956)

"Sometimes science is a lot more art than science. A lot of people don't get that."

—*Rick Potion #9*, written by Justin Roiland,
Rick and Morty 1.06 (2014)

TIME AS THE FOURTH DIMENSION

Sometimes it takes the art of science fiction to show science the way. Consider the dimensionality of reality, for example. Our world has four dimensions. Three dimensions are space; time is the fourth. Now, that may seem very obvious today, steeped as we are in the science-fictional soup that is modern culture. And yet, it was H.G. Wells who first introduced us to the idea of time as the fourth dimension in his classic 1895 novel, *The Time Machine*. Wells had his time traveler explain it in simple terms, "There is no difference between time and any of the three dimensions of space, except that our consciousness moves along it." This notion was nothing new. Relating time to space has a long history, going all the way back to ancient Greek thinker Aristotle. Industrialization led to a nineteenth-century fixation with time, but even then, most people thought of the fourth dimension as spatial. Wells begged to differ, and in so doing, he opened up a new and fascinating chapter in the evolution of ideas about time.

Time was in the ether. It splashed upon the canvas of Cubism, one of the most influential art movements of the twentieth century. Cubists such as Pablo Picasso and Georges Braque produced paintings where various viewpoints were visible in the same plane at the same time. The dimensions of space and time were used to give the artist's subject a more profound sense of depth. It was a revolutionary new way of looking at reality.

Time was captured in early cinema, too. The stop-motion photography of Étienne-Jules Marey inspired French artist Marcel Duchamp to paint his highly controversial *Nude Descending a Staircase*, which depicted time and motion by successive superimposed images. It also inspired Futurism, the most important Italian avant-garde art movement of the twentieth century. The Futurists celebrated technology and urban modernity and were fascinated by the challenge of rendering modern life. They strived in their paintings to evoke the modern experience, the contemporary human condition that embraced all kinds of sensations, and not just those visible to the eye. In their mission, the Futurists used time in the form of chronophotography, which allowed the movement of an object to be portrayed in time, not over a sequence of frames, but in a single frame of art.

Physics caught the time fever when Einstein introduced Special Relativity in 1905. The idea of time had changed over the ages before Einstein, as the quotes from great thinkers at the start of this chapter show. Lucretius suggested that time didn't exist by itself but was inextricably linked to reality, whereas Newton regarded time and space as a backdrop against which physical phenomena occur. At first, Einstein spoke of three spatial dimensions, and time. It was only after his teacher, Hermann Minkowski, promoted the view of time as the fourth dimension that the notion of the space-time continuum was created. It was essential to the development of Einstein's later work in General Relativity, and it is the same concept that Wells pioneered.

SPACETIME AND INTERSTELLAR

Spacetime was born. Einstein gave the world a new take on the fourth dimension. Moving clocks run slow, time is slowed down by gravity, and the speed of light is the same no matter how the observer is moving. It was another revolution in time. It was embraced by many but seemed to seriously worry Spanish surrealist Salvador Dalí, whose anxiety is palpable in his famous painting *The Persistence of Memory*. For many, Dalí's floppy clocks are history's most graphic illustration of the new playfulness of distorted Einsteinian time.

Einstein's new take on time was a gift to science fiction. One fascination was the time paradox, which can be nut-shelled by the question "What would happen if I went back in time and killed my own granddad?" The skill of tampering with time in this way reached its genius peak with Robert A. Heinlein's *All You Zombies* (1959), which also formed the basis of the 2014 Australian science-fiction film, *Predestination*. The main character of Heinlein's story moves along the fourth dimension, undergoes a sex change, and becomes his/her own mother and father. Weird.

A notable recent marriage of Einsteinian spacetime physics and science fiction is the 2014 movie *Interstellar*, winner of the 87th Academy Award for Best Visual Effects. The brain behind the science bits of the film belonged to Kip Thorne, an American theoretical physicist and Nobel laureate, well known for his contributions to astrophysics, and consultant to movies such as Carl Sagan's *Contact* in addition to Christopher Nolan's *Interstellar*. Under Thorne's guiding hand, *Interstellar* became another sterling example of how science fiction is able to work out highly complex physics in the form of a story, which we can better hook our brains around when we see it up on the screen. Having said that, the plotline is still a fascinating challenge to the senses.

In *Interstellar*, time warping comes about through black hole physics. Falling into the event horizon of a black hole named Gargantua, the film's protagonist, played by Matthew McConaughey, approaches a singularity or region in spacetime in which gravitational forces become infinite. The story features a "gentle singularity," the kind that doesn't kill, and McConaughey is scooped up by a tesseract or four-dimensional analogue of the cube that acts as a kind of space vehicle, possibly built by humans in the far future. The tesseract transports him to the higher dimension—from which McConaughey can see through the bookshelves of his daughter's room back on Earth—and across time so he is able to weakly interact with Earth's gravity and in so doing influence events there.

How could he have gotten back to his daughter's childhood? The movie's guiding scientific principle is that no human or object can go backward in time in our universe. However, a fifth dimension exists outside our universe. Humans cannot access this fifth dimension. Indeed, the only thing that *can* access the fifth dimension (and this is real physics borrowed from Einstein's general relativity theory carried into the fifth dimension) is gravity. McConaughey's character uses closed time-like curves in the fifth dimension to influence events back on Earth. His daughter, Murph, is in one face of the tesseract, and McConaughey is in another face of the tesseract. McConaughey pushes on a structure to produce a gravitational force that affects the interior of the tesseract, which is in the higher dimension, and that gravity force goes backward in time and influences events in his daughter's spacetime. If one should wonder at the complexity of the plot, Kip Thorne claims in his book *The Science of Interstellar* that Christopher Nolan always admired Stanley Kubrick's *2001*, and especially the ending, which is just as tricky to understand.

After all that, and looking for a little light relief, it is worth remembering the clever use of dimensions in Kurt Vonnegut's 1973 novel, *Breakfast of Champions*. Though we live in a three-dimensional world, Vonnegut reminds us, there may well be other dimensions we cannot perceive. It was on this basis that a character in the book was described as having "a penis eight-hundred-miles long and two hundred and ten miles in diameter, but practically all of it was in the fourth dimension."

LOOPER:

WILL WE EVER BUILD A TIME MACHINE?

"Long ago I had a vague inkling of a machine . . . that shall travel indifferently in any direction of Space and Time, as the driver determines."
—The time traveler in H.G. Wells's, *The Time Machine* (1895)

"That is the germ of my great discovery. But you are wrong to say that we cannot move about in Time. For instance, if I am recalling an incident very vividly I go back to the instant of its occurrence: I become absent-minded, as you say. I jump back for a moment. Of course, we have no means of staying back for any length of Time, any more than a savage or an animal has of staying six feet above the ground. But a civilized man is better off than the savage in this respect. He can go up against gravitation in a balloon, and why should he not hope that ultimately, he may be able to stop or accelerate his drift along the Time-Dimension, or even turnabout and travel the other way?
—The time traveler in H.G. Wells's, *The Time Machine* (1895)

"Time travel has not yet been invented. But thirty years from now, it will have been. It will be instantly outlawed, used only in secret by the largest criminal organizations. It's nearly impossible to dispose of a body in the future . . . I'm told. Tagging techniques, whatnot. So, when these criminal organizations in the future need someone gone, they use specialized assassins in our present called 'loopers.' And so, my employers in the future nab the target, they zap him back to me, their looper. He appears, hands tied and head sacked, and I do the necessaries. Collect my silver. So, the target is vanished from the future, and I've just disposed of a body that technically does not exist. Clean."

—*Looper*, screenplay by Rian Johnson (2012)

TIME MACHINES

Your machine is made. Where in time do you first travel? What momentous event in world history would make your list? Columbus first setting foot on the Americas in 1492, perhaps? Your timely intervention might save a few million lives. Or maybe you journey to the prehistoric past on the eve of the demise of the dinosaurs to see the mother of all extinctions unfold at first hand before beating a hasty retreat in your machine. Or maybe you travel back in time to 117 AD, when the grand Roman Empire under Emperor Trajan had reached its greatest height. Death of the dinosaurs, you say? Good choice.

Prime your machine. Temporal crankshaft. Check. Fourth-dimensional perambulator. Check. Ignition. Check. Ignition? Wait, what? Exactly how are these time machines actually meant to work?

One of the very first temporal gadgets was "The Outlandish Watch." It was invented by Charles Dodgson, better known to the world as Lewis Carroll. Dodgson's tiny time machine appeared in the 1889 story *Sylvie and Bruno*, which he wrote under the name of Lewis Carroll. The watch had two modes. If the reverse peg was pushed, "events of the next hour happen in reverse order." The other mode involved the watch's hands. They could be moved backward, as much as a month, enabling the wearer of the watch to travel into the past.

Naturally, you're not going to find one of these bad boys at your local jeweler. The watch's owner, a professor, for heaven's sake, is frustratingly elusive when asked to describe the theoretical basis for the thingamajig: "I could explain it, but you would not understand it." So much for science communication. Sadly, this kind of irresponsibility is common with owners of time machines. Indeed, Dodgson wasn't even the first inventor of a fictional time machine. Two other writers had beaten him to it. US newspaperman and author Edward Page Mitchell was the first to invent such a gadget in his 1881 story, *The Clock That Went Backward*, a tale that may well have influenced Wells himself. Spanish diplomat Enrique Gaspar then wrote *El Anacronópete* in 1887, which described a time machine as an elaborate and hermetically sealed ark whose lavish furnishings and plenteous passenger list are similar to the Nautilus in Jules Verne's 1871 story, *Twenty Thousand Leagues Under the Sea*. These early time-machine stories were rich in clocks, watches, and lush furniture but offered little detail of the time mechanics.

H.G. Wells made little more effort. He shrewdly made the description of his time machine extremely vague. It was left to others to figure out. French writer Alfred Jarry stuck his neck out in a review of Wells's 1895 book, *The Time Machine*. The review was pithily titled *A Commentary to Serve for the Practical Construction of the Machine to Explore Time*. Wells had talked vaguely of levers, but Jarry unduly designed his machine with three rotating gyrostats. He even included an impressive nonsense diagram. Jarry figured that a time machine would have to anchor itself absolutely in space in order to move in time. In this way, he says, "all future and past instants . . . would be explored successively."

Incidentally, when MGM came to make Wells's time machine for the 1960 movie, it was a steampunk dream. The iconic *Time Machine* prop was codesigned by director George Pal and MGM art director William Ferrari. Pal plumped for the look of a horse-drawn sled, inspired by winter sled rides of his boyhood. The machine was made with a traditional barber's chair, which Pal picked, as it was reminiscent of the kind of pilot's seat a time traveler might need. Pal produced what many believe to be the best and most elegant time contraption—all brass, glass, and dials. With its red leather padded seat, it's exactly the kind of beast you imagine a Victorian time-traveling gent *would* "drive." Sadly, this wonderful steampunk creation, built by designer Wah Chang, was destroyed in a fire at George Pal's Bel-Air house in November 1961.

Talk of iconic time machines must surely bring us to the TARDIS, the Time and Relative Dimension in Space machine, in *Doctor Who*. At first glance, the TARDIS is a fairly faithful fit of the police telephone boxes that could be seen on the streets of Britain in the sixties. The explanation of the appearance of the TARDIS is ingenious. The ship sported a "chameleon circuit," a mechanism responsible for changing the outside appearance of the ship the millisecond it lands in order to fit in with its environment, but the circuit broke, fixing the TARDIS as a phone box. As the TARDIS is said to have traveled over 100 trillion years, from before the Big Bang to the edge of time itself, it must surely be one of the most well-used time machines. No doubt the TARDIS was the main influence behind the Phone Booth from the comic time-traveling 1988 movie, *Bill and Ted's Excellent Adventure*. Whereas the Phone Booth didn't have the spatial quirkiness of the TARDIS, totally lacking its bigger-on-the-inside facility, the tale did come with perhaps the best catchphrase in time-machine culture with "Gentlemen, we're history!"

GETTING JIGGY WITH TIME

Looking for the latest in gadgetry convenience, some science-fiction writers of the 1960s and '70s then went after wearable time tech, with the hot favorite being the time belt. Apple, eat your heart out. The last of these time-travel belts even allowed sideways travel into alternate realities, but no doubt at an extra cost. In Bob Shaw's 1977 story, *Who Goes There?*, the time machine is a small cage-like construction in which "the rods meet at strange angles and create a wrenching sensation in my eyes." Alas, the device met its end when it became irreparably bent out of shape because someone "sat on it last week."

Kurt Vonnegut's race of fictional aliens, the Tralfamadorians, had gone even further. With no gadget of any kind, they were naturally able to see along the time line of the universe. They not only had the ability to experience all four dimensions, they also had total recall of both past and future. The human view of time is a mere snapshot. The Tralfamadorian's is that of a movie, in which all scenes are played out at once. This Tralfamadorian view of time even seemed to include Albert Einstein. He wrote in 1955 that relativists understood the division of past, present, and future to be an illusion.

The 1995 movie *Twelve Monkeys* featured a "projector-collector" contraption, which enabled Bruce Willis to be pinged back in time to trace the outbreak of a deadly virus. The trouble was, it seemed the machine wouldn't work unless the time traveler was naked, which might prove tricky in certain social situations. Curiously, naked time travel also seemed to be the norm in the trilogy of *Terminator* movies. Though we seldom see the actual machine, we do learn that it only seems to send its traveler one-way in time.

In the 2004 movie *The Butterfly Effect*, travelers were transported back in time by diary. No doubt latching onto the idea of how the human mind is often imaginatively transported into history when reading a book, or looking at a work of art, the travelers in *The Butterfly Effect* read their own words and are transported back to the time when the words were written. Are written. As usual, it gets complicated, as is the case of the time machine in the alternative history television series, *Quantum Leap*. The time machine in this series merely seemed to be a puff of smoke in a darkened but futuristic-looking room. Complications arose from the fact that you only travel within your own lifetime and could end up in other people's bodies, which some might find a little gross.

More than a century of science-fictional talk of time machines has influenced scientists to ask how feasible time machine trips might be in actual fact. Professor of Theoretical Physics Paul Davies addressed such questions in his 2001 book, *How to Build a Time Machine*. If we want to travel into the future, all we need is a machine that can move at a velocity close to the speed of light. As our spaceship approaches this speed, the slower time moves. Once you get back to Earth, you will hardly have aged. Decades, or even centuries, will have passed "back home."

According to physicist J. Richard Gott in his 2002 book, *Time Travel in Einstein's Universe*, traveling back in time is far trickier. It entails fiddling with wormholes, cosmic strings, or black holes—the kind of time machines made feasible only with mind-warping technology. In 2007, American scientist Ronald Mallett broke the news of his life-long struggle to build a time machine. Mallett's take on temporal travel is to bend spacetime. Massive objects such as stars and planets do this. Mallett is among those who believe that light too can bend the continuum. Rather than the De Lorean envisaged in *Back to the Future* films or the machine from the 2012 movie *Looper*, which looks a lot like a space capsule with a bunch of spaghetti wiring and two huge batteries fixed to the door, Mallett's machine is a ring laser—an extremely powerful one. Mallett hopes that, one day, by simply popping into this huge vortex of light, travel through time might be possible.

2001: A SPACE ODYSSEY:

IS THERE EVIDENCE OF GUIDED EVOLUTION IN HUMAN HISTORY?

KUBRICK: "I will say that the God concept is at the heart of 2001—but not any traditional, anthropomorphic image of God. I don't believe in any of Earth's monotheistic religions, but I do believe that one can construct an intriguing scientific definition of God, once you accept the fact that there are approximately 100 billion stars in our galaxy alone, that its star is a life-giving sun and that there are approximately 100 billion galaxies in just the visible universe . . . so it seems likely that there are billions of planets in the universe not only where intelligent life is on a lower scale than man but other billions where it is approximately equal and others still where it is hundreds of thousands of years in advance of us."

PLAYBOY: "If such creatures do exist, why should they be interested in man?"

KUBRICK: "They may not be. But why should man be interested in microbes? The motives of such beings would be as alien to us as their intelligence."

—Stanley Kubrick, *Playboy* (1968)

CARAS: "In all your experience, is there any evidence at all, anywhere, that either a culture has been influenced by an extraterrestrial visitation, or that there has been anything that even resembles an artifact?"

MEAD: "Some people have claimed that the only way we can explain some of the high civilizations of the past is by extraterrestrial life having landed here at some point and taught everybody, everything in things like Stonehenge for instance and the complexity of the astronomical measurements that they have found when they looked at it. Now this relates to a lot of our other attempts—in the past—to think of only one bright civilization somewhere and give it the responsibility for everything else."

—Interview between Roger Caras and anthropologist Margaret Mead in *Are We Alone? The Stanley Kubrick Extraterrestrial Intelligence Interviews* (2005)

"Some years ago, I came upon a legend, which more nearly fulfils some of our criteria for a genuine contact myth. It is of special interest because it relates to the origin of Sumerian civilization. Sumer was an early—perhaps the first—civilization in the contemporary sense on the planet Earth. We do not know where the Sumerians came from. Their language was strange; it had no cognates with any known Indo-European, Semitic, or other language, and is understood only because a later people, the Akkadians, compiled extensive Sumerian-Akkadian dictionaries."

—Carl Sagan and Iosif Shklovsky, *Intelligent Life in the Universe* (1966)

"The baby's blood type? Human, mostly."

—Orson Scott Card, *Six Word Story* (2006)

GUIDED EVOLUTION

"A scientific definition of God." That's how Stanley Kubrick described his 1968 movie masterpiece, *2001: A Space Odyssey*. Roughly eighty-four years earlier, Darwin had stirred German thinker Friedrich Nietzsche to recognize three phases in the evolution of humans: ape, man, and finally, superman. As Nietzsche said in his book *Thus Spoke Zarathustra*, "What is the ape to man? A laughingstock, or painful embarrassment. And man shall be to the superman: a laughingstock or a painful embarrassment." For Nietzsche, modern humans were solely the stepping-stone from ape to superman.

The trouble is, neither Darwin nor Nietzsche are exactly primetime TV. There's little drama in Darwinian evolution, save the odd cosmic catastrophe every million years or so. What remains is the sluggish, unsolicited course of creeping change. So, along with famous British science-fiction author Arthur C. Clarke, Kubrick tried turbocharging Nietzsche and souping-up Darwin: together, Clarke and Kubrick developed a fictional form of Stephen J. Gould's "punctuated equilibrium," those rare times when swift and crucial evolutionary change happens. In *2001*, dumb, blind evolution is ruptured by the periodic guiding hand of an elusive alien race. It is a story of the effective creation and resurrection of man in three acts. A scientific definition of God, as Kubrick called it.

The brilliant use of movie technology helped make *2001* a tour de force. Winner of an Academy Award for special effects, the film was said by some to present an even more "realistic" portrait of space travel than the real-life odyssey of Armstrong and Aldrin a year later. Sections of the film were used in training NASA astronauts, and Arthur C. Clarke suggested that of all the reactions to the movie, he was most pleased by the response of Soviet cosmonaut Alexei Leonov: "Now I feel I've been into space *twice!*"

As the title of Kubrick's movie implies, the story takes the viewer on a journey in time and space. The message of the movie is not explicit. As Kubrick told *Playboy* in an interview in 1968, "*2001* is a nonverbal experience; out of two hours and nineteen minutes of film, there are only a little less than forty minutes of dialog. I tried to create a visual experience, one that bypasses verbalized pigeonholing and directly penetrates the subconscious with an emotional and philosophic content . . . in *2001* the message is the medium." Though Kubrick told *Playboy* the audience member should be "free to speculate as you wish about the philosophical and allegorical meaning of the film," he nonetheless hoped that the movie would succeed in making the viewer give thought to "man's destiny, his role in the cosmos and his relationship to higher forms of life."

2001 is an immense four-million-year filmic story. The story that unfolds starts with a subhuman ape and ends with the posthuman star-child. Many science fiction films focus on just a single theme, but *2001* has them all. As the movie unravels, each theme of science fiction is revealed: the *space* theme in the form of alien contact through the monoliths; the *time* dynamic, as *2001* is essentially an evolutionary fable; the *machine* motif, especially through the presence of HAL, the famous computer-turned-murderer; and the *monster* theme, speculating what man may one day become with the movie's metamorphosis of modern humans into supermen.

In the opening "Dawn of Man" scene of *2001*, we see the sunrise above the primeval plains of Earth and listen to the rousing soundtrack of Richard Strauss's Nietzsche-inspired tone poem, *Also Sprach Zarathustra*. A small band of man-apes is on the long, pathetic road to racial extinction. The journey begins with one of the hominids proudly hurling an animal bone into the air. In an astounding cinematic ellipsis, the bone instantly morphs into an orbiting satellite, and three million years of hominid evolution is written off in one frame of film. (It's one of the most famous "match cuts" in all of cinema.) The agency that drives the guided evolution of these early hominids is an alien artifact in the shape of a black monolith. Like the Martians in H.G. Wells's *The War of the Worlds*, the monolith embodies the void. Primal bone technology marks the birth of the modern era, in which man and machine, from the very outset, are inseparable. The mysterious presence of the monolith transforms the hominid horizon. The journey to superman begins.

In that single frame of match-cut film, the space age dawns. It is a bland future, dominated by corporations and technology. It's as if Kubrick were mapping our presence in the story's journey and saying, "YOU ARE HERE." Ironically, the most "human" character is the robustly intelligent spaceship computer, HAL 9000. The potent evolutionary force imparted by the black obelisks is once more overdue.

The space age was originally inspired out of the apes by an alien intelligence in the form of the monoliths. Once more, the odyssey of human self-discovery culminates under the watchful presence of the monoliths when modern man, in the form of the individual astronaut David Bowman, comes to an end. With the movie screen replete with the massive presence of planet Earth, the fetus of the superhuman star-child floats into view. The star-child moves through spacetime without artifice, the image suggesting a new power. Man has transcended all earthly limitations.

KUBRICK'S BIBLE

2001 had been many years in the making. Kubrick's bible for the film was *Intelligent Life in the Universe*, written in 1966 by American astronomer Carl Sagan and Russian astrophysicist Iosif Shklovsky. The book was the first seminal scientific text on the question of extraterrestrial intelligence. (Kubrick had originally filmed interviews with twenty-one leading scientists to act as a prologue to the film's main narrative. Interviewees included physicists Frank Drake and Freeman Dyson, anthropologist Margaret Mead, roboticist Marvin Minsky, and Alexander Oparin, the great Soviet authority on the origin of life, sometimes described as the "Darwin of the twentieth century." Kubrick's aim was to lend astrobiology that special dignity it has only acquired since. Though the interviews were cut from the final version of the film, a book of the transcripts was published in 2005.)

Sagan and Shklovsky had also speculated about guided evolution in *Intelligent Life in the Universe*. They based their argument on the Drake equation, a calculation of how many communicating alien civilizations there might be in our Milky Way galaxy. Their reasoning went something like this: if there are sufficient technically accomplished alien civilizations out in the galaxy, then some will look for contact with other less-developed planets and every 100,000 years or so visit these planets to guide and monitor their evolution. Should this monitoring be on a more frequent basis, say, every several thousand years, then alien contact may have occurred within historical times, and evidence might exist if you knew what you were looking for.

To illustrate their point, Sagan and Shklovsky pondered the incredible origin of the Sumerian civilization. Making it clear that they were not claiming "an example of extraterrestrial contact," the two scientists nevertheless concluded that "the legend suggests that contact between human beings and a nonhuman civilization of immense powers on the shores of the Persian Gulf" had occurred somewhere in Sumer in the fourth millennium BC or earlier. Using a number of independent historical accounts from ancient writers, Sumerian civilization is presented by the descendants of the Sumerians themselves to be of nonhuman origin. Their accounts detail a succession of strange creatures, which appear over the course of several generations. These accounts speculate that the sole purpose of the strange creatures is to educate and guide humans. Each is aware of the mission and achievements of his predecessor, and when a great flood threatens the newly acquired knowledge that humans have, steps are taken to safeguard its preservation. As Sagan and Shklovsky note, "The Straightforward nature of this account of contact with superior beings is notable. [The visitors] are described variously as 'animals endowed with reason,' as 'beings,' as 'semi-daemons,' and as 'personages.' They are never described as gods."

Only time will tell whether we find hard evidence of guided evolution in our past. However, Hungarian linguist and systems scientist Béla Bánáthy is more interested in what humans may one day become. In his book *Guided Evolution of Society*, Bánáthy suggests the first decisive episode occurred several million years ago, when our hominid ancestors first developed. The second crucial event happened when Homo sapiens began the revolutionary route of cultural evolution. And today, says Bánáthy, we have arrived at the third major event: the revolution of conscious evolution and the power and responsibility we possess to guide the evolutionary journey of our species.

WHAT'S THE HISTORY OF SCIENCE FICTION IN SEVEN OBJECTS?

"We travel back in time and across the globe, to see how we humans have shaped our world and been shaped by it over the past two million years. The book tries to tell a history of the world ... by deciphering the messages which objects communicate across time. . . . They speak of whole societies and complex processes rather than individual events, and tell of the world for which they were made, as well as of the later periods which reshaped and relocated them. . . . It is the things humanity has made . . . and their often curious journeys across centuries . . . which *A History of the World in 100 Objects* tries to bring to life.

—Neil MacGregor, *Signals from the Past,* from an introduction
to *A History of the World in 100 Objects* (2010)

Signals from the Past

In 2010, the British Museum and the BBC decided to grapple with history. After four years of planning and presenting over a period of six months on BBC Radio 4, they detailed 100 objects of past art and technology, all of which are in the British Museum, that tell a history of the world. The 100 objects ranged from a Tanzanian million-year-old hand axe and the statue of a Minoan Bull-leaper from the ancient culture on Crete, to the ship's chronometer from Darwin's HMS *Beagle.* The Museum suggested that this landmark project told *a* history, not *the* history. So, what better way to tell the history of science fiction on modern culture than to set up our own experiment? For want of space, we shall reduce our list from 100 objects down to a magical seven, but we will retain the same intention to briefly tell the chronological tale of each artifact from *a* chosen history of science fiction.

1: False Maria

The sleek, metallic, and arty-looking female robot from the 1927 movie, *Metropolis*. Fritz Lang's *Metropolis* was a science-fiction cinematic extravaganza, produced in Germany at the height of the Weimar Republic, and the most expensive silent movie of its day. Lang's stylistic and seminal work has been dubbed "raygun gothic." The film featured an architecture based on contemporary Modernism and Art Deco, but it was also a social comment on design ideas for the modern environment, a futuristic dystopia of skyscrapers and social class.

False Maria is not just an inhabitant of the futuristic city of Metropolis of the year 2026, she is one of the first mechanized humans of the movies. She danced topless. She sparked duels. She used her charismatic powers to entrance the populace of Lang's dystopian society. She incited riots, as her one true intention was overthrowing the ruling class. She was a revolutionary robot! Little wonder she inspired every robot vision for the next century. Rumors that she is the mom of C3PO are greatly exaggerated.

2: The Cape of Superman

Ever since his first incarnation in *Action Comics 1* of June 1938, for many people Superman has been, and still remains, the most iconic superhero of them all. Not only does Kal-El boast strength, speed, and flight among a sheer number of impressive powers, but he's also from another planet. His many adventures include *Red Son* by Mark Millar, an alternate history where Superman crash-lands in Russia during the days of Joseph Stalin and becomes a communist leader, and a recent tale in which he renounces his American citizenship to better serve the whole of planet Earth.

Which single artifact epitomizes the Man of Steel and the hordes of superheroes that were to follow? Superman's red cape. For in almost all versions, Superman's cape is adorned with the "S" shield, representing one of the first superheroes and serving as a template for superhero character design decades after Kal-El's first appearance. Indeed, the lore of wearing a characteristic symbol, usually on the chest, was copied by many subsequent superheroes, including Batman, Spiderman, Wonder Woman, and the Flash. Historians in the future might variously interpret the meaning of that "S" as a family crest, a snake, the Kryptonian symbol for hope, or just plain Superman himself.

3: The TARDIS

The world's most famous space-and-time machine, from *Doctor Who*. Owing much to Albert Einstein and his theory of relativity, the TARDIS stands for Time and Relative Dimensions in Space. Designed in 1963, it's famously bigger on the inside and boasts a chameleon circuit, which allows it to blend with its immediate surroundings, wherever and whenever it lands. The circuit broke early on, resolving the chameleon into a phone box so iconic that the design is now more associated with the fictional TARDIS than the original police box.

The police box design is trademarked by the BBC, despite the fact that the design was originally created by the London Metropolitan Police. The word TARDIS is also an entry in the Oxford English Dictionary. Indeed, the TARDIS is so iconic, it is featured in marketing pitches by Apple and Google, as well as countless cultural references in all forms of media. (An honorable time-machine reference here should also go out to the DeLorean from the *Back to the Future* franchise, another fictional time-machine design more famous than its original factual counterpart.)

4: The Ansible

The ansible was a device capable of instantaneous or faster-than-light communication. It can send and receive communication to and from other ansibles over any distance in spacetime, and without delay. The word was coined in 1966 by American writer Ursula K. Le Guin but has since been shared across a number of stories and settings, including both the book and movie of Orson Scott Card's *Ender's Game*. Indeed, though not named in the movies, the kind of coordinated space battles seen in *Star Wars* and *Battlestar Galactica* would not be feasible without an ansible.

5: The Voight-Kampff Machine

American science-fiction writer Philip K. Dick amassed an immense volume of work spanning forty-four books and hundreds of short stories. He was merely on the brink of worldwide success when he died aged fifty-four, success that was going to be created largely through the big screen adaptation of his 1968 novel, *Do Androids Dream of Electric Sheep?* Dick's book was transformed into *Blade Runner* by director Ridley Scott. Hugely influential in both its themes and cinematography, *Blade Runner* became a pivotal part of the new science-fictional movement that was to encapsulate the scientific advances of the late twentieth century—cyberpunk.

Originating in Dick's book, the Voight-Kampff machine is a fictional polygraph-type contraption, used by the LAPD, to test whether a subject is an android replicant or human. The machine measures bodily data such as respiration, heart rate, blush level, and eye movement, in response to emotionally provocative questions from the tester. *Blade Runner* actually begins with a Voight-Kampff test. Key to the book, though not so prominent in the movie, was Dick's contention that empathy was a defining human trait.

6: HAL 9000

The main antagonist in Arthur C. Clarke's *Space Odyssey* series, HAL (**H**euristically programmed **AL**gorithmic computer) is most famous for appearing in Kubrick's *2001: A Space Odyssey.* Some of HAL's hardware can be spotted in the movie, but for the most part, all we see is a camera lens containing a red light and a yellow dot. Speaking in a soft and rational but sometimes menacing voice, the robust intelligence of HAL contrasts sharply and ironically with the banality and vacuity of the human crew members. HAL is cinema's first and most murderous AI. Listed as the 13th-greatest-ever movie villain in the American Film Institute's *100 Years Heroes & Villains*, HAL also has the 9,000th asteroid named after him, which orbits in the Belt somewhere between Mars and Jupiter.

7: The Lightsaber

The fact that the lightsaber needs no introduction simply shows how iconic this most famous of all science-fiction weapons is. (An honorable mention here should go to the Death Star, except any curator would find fitting the Death Star in a museum something of a cosmic challenge.) Indeed, the lightsaber is so famous that the main question is, Why didn't anyone think of it before George Lucas? The lightsaber brings back days of medieval chivalry, as in truth it's a high-tech update of the sword. The Jedi are even referred to as knights, and no doubt there's a round table in there, too, if you look hard enough. Whereas the range of a sword can be limited, the skill of a Jedi force-user makes a lightsaber the deadliest of weapons. With its distinctive hum, which rises in pitch and volume as the blade is moved rapidly through the air, the lightsaber is used by the Jedi for knowledge and defense, never for attack. In 2008, the *Daily Telegraph* reported a poll of over 200 movie fans in which the lightsaber was found to be the most popular weapon in film history.

MAD MAX:

IS SOCIETY RUNNING DOWN INTO CHAOS?

"People sometimes think technology just automatically gets better every year, but it actually doesn't. It only gets better if smart people work like crazy to make it better. That's how any technology actually gets better. Look at the history of civilization, many civilizations. And look at say ancient Egypt where they built all these incredible pyramids, and then they forgot basically how to build pyramids. And even hieroglyphics, they forgot how to read hieroglyphics. Or you look at Rome, and how they built these incredible road ways and aqueducts and indoor plumbing, they forgot how to do all of those things. And there are many such examples in history, so I think you should always bear in mind that entropy is not on your side."

—Elon Musk, podcast *How to Build the Future* (2016)

"A good many times I have been present at gatherings of people who, by the standards of the traditional culture, are thought highly educated and who have with considerable gusto been expressing their incredulity at the illiteracy of scientists. Once or twice I have been provoked and have asked the company how many of them could describe the Second Law of Thermodynamics. The response was cold: it was also negative. Yet I was asking something which is about the scientific equivalent of: Have you read a work of Shakespeare's?"

—C. P. Snow, *The Two Cultures and the Scientific Revolution* (1959)

"The law that entropy always increases, holds, I think, the supreme position among the laws of Nature . . . if your theory is found to be against the second law of thermodynamics I can give you no hope; there is nothing for it but to collapse in deepest humiliation."

—Arthur Eddington, *The Nature of the Physical World* (1928)

"As for man, there is little reason to think that he can in the long run escape the fate of other creatures, and if there is a biological law of flux and reflux, his situation is now a highly perilous one. During ten thousand years his numbers have been on the upgrade in spite of wars, pestilences, and famines. This increase in population has become more and more rapid. Biologically, man has for too long a time been rolling an uninterrupted run of sevens."

—George R. Stewart, *Earth Abides* (1949)

"I know not with what weapons World War III will be fought, but World War IV will be fought with sticks and stones."

—Albert Einstein, interview with Alfred Werner,
Liberal Judaism 16 (April–May 1949)

ENTROPY

Picture yourself on a terminal beach. You had set the controls of your time machine to the max. Now you find yourself in the foggy ruins of the far future. You've journeyed to the end of time. The Earth is locked by tidal forces. The planet spirals toward a red giant Sun, which hangs motionless in an endless sunset. The solar system is in meltdown. So ends H.G. Wells's prophetic and terrible account of the end of the world in his famous 1895 novel, *The Time Machine*. Wells took a momentous leap with his bleak portrayal of Darwinian evolution. He depicted the entire cosmos as a machine running low on energy. The universe is in a state of entropic decay, and man is being swept away "into the darkness from which his universe arose."

With the development of cosmology in the 1960s, astronomers realized Wells may have been right. This image of a dying cosmos and a drift in entropy became known as "the heat death of the universe." When science fiction entered a new and more daring age in the swinging sixties, the scenario for the heat death moved from the cosmic to the personal, and from outer space to inner space. Polymath Pamela Zoline's *Heat Death of the Universe* was published in 1967 and is one of the most underappreciated works of science fiction. The story consists of fifty-four numbered paragraphs describing the domestic drudgery of Sarah Boyle, a Californian "mom" as she prepares for her child's birthday. Interlaced with this familial narrative are reflections on entropy, chaos, and the heat death of the universe itself. *Heat Death of the Universe* was a modern classic of the new science fiction, with its incongruity, experimentalism, and political overtones. Zoline wrote the story for *New Worlds* magazine, edited by British writer Michael Moorcock and increasingly featuring experimental and avant-garde material of science fiction's radical "new wave." Moorcock, a very able writer in his own right, commended Zoline's short story for the way in which it connected the modern myths of science such as entropy with the urgent theme of modern fiction—the victimized housewife. The subversion of domestic ideals placed wryly alongside the gradual exhaustion of the entire cosmos fused the mundane with the cosmic. Zoline's story, like many tales of the time, embraced new patterns of thinking, feeling, and behaving in a world where students and young people began to break free from tradition and welcome a new orthodoxy that proclaimed, *Tune in, turn on, and drop out!*

Since Hiroshima and Nagasaki, science fiction had been acclimatizing us to a mood of apocalyptic terror. Millions of people responded anxiously to the catastrophic visions of an entropic future. The impact of the atomic bomb was clear. The future wasn't what it used to be. Gone were the monorails, the silver suits, and the generation starships of past fiction. In their place were catastrophe and the ominous image of the thermonuclear mushroom.

There'd been apocalyptic fiction before the bomb. Eight years after *Frankenstein*, Mary Shelley had published *The Last Man*, the first novel to describe the extinction of humans. Her book ends at the close of the twenty-first century, when, after societal progress, war unleashes a worldwide plague. The last survivor roams a desolate planet in a futile search for another living soul.

Another "last man" novel was the 1901 book *The Purple Cloud* by M.P. Shiel. Described as brilliant by H.G. Wells and as exemplary weird fiction by H.P. Lovecraft, Shiel's tale featured a race to the North Pole. The protagonist, Adam Jeffson, inadvertently escapes a worldwide disaster by which a gaseous purple cloud destroys almost all global fauna, leaving him as the sole survivor of a ravaged earth. Despite the prose at times being as purple as the eponymous cloud, the scenes of isolation are convincingly done, and, as H.P. Lovecraft says in his *Supernatural Horror in Literature*, the story is "delivered with a skill and artistry falling little short of actual majesty."

EARTH ABIDES

A society running down into chaos was also the theme of George R. Stewart's *Earth Abides* (1949). Here, the catastrophe is a global disease with no cure. Stewart shuns the atomic obsession of the age; he wipes out most of humanity with a plague. Only those naturally resistant, or geographically isolated, survive. In this primitive apocalypse of man versus nature, Stewart is ambivalent as to the cause. The devastation may have been triggered by human error, an "escape, possibly even a vindictive release, from some laboratory of bacteriological warfare." Or perhaps it is a natural process of decay caused by the "biological law of flux and reflux."

The main character in *Earth Abides*, Ish, emerges from the hills into an alien world. Overnight, modern society has been transformed into an agrarian postapocalyptic landscape. The book provides great contrast to the brazen survivalist accounts that imagine civilization can be re-established in a few weeks with a semiwrecked vehicle, a hacksaw, and a Swiss army knife. *Earth Abides* suggests that civilization may never be the same again, the novel's title echoing the Book of Ecclesiastes, "Men go and come, but Earth abides."

The striking aspect of *Earth Abides* is its lack of postapocalyptic clichés. There's no shortage of shelter or rations. Absent are the leprous biker gangs and the requisite roving mob of mercenaries. Nor is there a farcical final battle between good and evil. Instead, Ish is a scientist antihero; low on survival skills, high on brains. The story recognizes that even scientists are laden with human frailties, unlike the archetypal heroes of pulp fiction.

Earth Abides chronicles Ish's postapocalyptic life. His main effort is to restore society among the small band of plague survivors. The story focuses on the lasting positive effect on the ecosystem once the blight of industry has vanished. Indeed, at one stage, Ish sees a flickering *Coca-Cola* sign in the distance and wonders how long the grid will keep alive this suspect symbol of civilization.

Ish is the last emissary of American civilization. The last American. *Earth Abides*, like many postapocalyptic stories, appealed to commonly held secret desires. Sanctuary from the confines of civilized society, a less populated world, and the chance to test one's mettle against the elements.

Albert Einstein is quoted in *Einstein's Mirror*, written by Hey and Walters, as saying in 1945, "I do not believe that civilization will be wiped out in a war fought with the atomic bomb. Perhaps two-thirds of the people of the Earth might be killed. But enough men capable of thinking, and enough books, would be left to start again, and civilization could be restored." In contrast, *Earth Abides* is an elegy to mankind. By the story's end, the community Ish founded has grown into a motley crew of superstitious hunter-gatherers, as "primitive" as the Neanderthal, and totally uninterested in rebuilding "civilization."

Such was the long road that led to *Mad Max: Fury Road*. In this 2015 movie, to some extent a reboot of the *Mad Max* franchise that began in 1979, pockets of civilization struggle to survive after the world has become a desert wasteland in the wake of a nuclear holocaust. Indeed, all this science-fiction talk about the entropic decay of societies led American academic Jared Diamond to consider the facts of the matter. In his 2005 book, *Collapse: How Societies Choose to Fail or Succeed*, Diamond looks at the causes of historical examples of the collapse of a number of societies. His conclusion is an interesting one for fans of *Mad Max*: "In fact, one of the main lessons to be learned from the collapses of the Maya, Anasazi, Easter Islanders, and those other past societies . . . is that a society's steep decline may begin only a decade or two after the society reaches its peak numbers, wealth, and power." Buckle up your War Rig, Dorothy, 'cause "Kansas" may be going bye-bye.

THE MINORITY REPORT AND THE MATRIX:
WILL WE EVER DEVELOP PRECOGNITION?

"Those who have knowledge, don't predict. Those who predict, don't have knowledge."
> —Lao-tzu, Chinese poet and scholar (6th century BC)

"The lily shall remain in a merry world; and he shall be moved against the seed of the lion and shall stand on one side of the country with a number of ships. Then shall the Son of Man, having a fierce beast in his arms, whose Kingdom is the land of the moon, which is dreaded throughout the world. With a number shall he pass many waters, and shall come to the land of the lion, looking for help from the beast of his country, and an eagle shall come out of the east, spread with the beams of the Son of Man, and shall destroy castles of the Thames. And there shall be battles among many kingdoms. That year shall be the bloody field, and lily F.K. shall lose his crown, and therewith shall be crowned the Son of Man K.W., and the fourth year shall be preferred. And there shall be a universal peace over the whole world, and there shall be plenty of fruits; and then he shall go to the land of the Cross."
> —Mother Shipton, inadvertently illustrating the bizarre
> and surreal nature of medieval prediction in *The Strange and
> Wonderful History of Mother Shipton* (1686)

"Prediction is very difficult, especially if it's about the future."
> —Niels Bohr, quoted in Arthur K. Ellis's *Teaching and
> Learning Elementary Social Studies* (1970)

"I never think of the future, it comes soon enough."
—Albert Einstein, quoted in David P. Sentner's article
"Would Take Three Days to Make Simple Explanation of
Theories" in Clearfield Progress (1930)

SOOTHSAYER

Prophecy has a long history long before the days of science fiction. All known ancient cultures had prophets who would deliver a prophecy as a message, which they claimed had come straight from the gods themselves. Typically, the message would reveal divine will on the prophet's social world, from which the prophet could, no doubt, profit.

Consider the more recent and curious case of old Mother Shipton. Location: Yorkshire, England. Date: said to be between the years of 1488 and 1561. Mother Shipton, real name Ursula Southell or Ursula Soothtell, was an infamous English soothsayer. She was a legend in her own lifetime, though her prophecies weren't published until 1641, eight decades after her death. A caricature of Mother Shipton was used in early pantomime, the forerunner of the ugly sisters of Cinderella, for example, and a species of moth. (The Mother Shipton Moth sports wing patterns that resemble a hag's head in profile.)

It wasn't until the 1684 edition of her prophecies that the world got its first published glimpse of the woman behind the forecasts. Said to be hideously ugly, Ursula was born in Knaresborough, a village in Yorkshire, and lived in a cave next to a Petrifying Well whose waters were so magicked that an object placed within their depths and left for a period of time acquired a stone-like appearance (an entirely natural process of evaporation and deposition in waters known to have a high mineral content, but weird enough to spook medieval minds). Predictably, the "Mother Shipton" cave and well are now tourist attractions, along with associated parkland, if you fancy a visit.

Known for predictions throughout her long life, many of Mother Shipton's forecasts were local and mundane (you know the kind of thing, whether the local farmer's crop yield will be good this year, or whether the vicar's wife has genital warts), yet she was also known for two prophetic well-known verses: "A carriage without horse will go, disaster fill the world with woe. In London, Primrose Hill shall be, in center hold a bishop's sea" is taken to predict not just the advent of the car and train, but more specifically the formation of the Church of England; "Around the world men's thoughts will fly, quick as the twinkling of an eye. And water shall great wonders do, how strange, and yet it shall come true" is meant to predict not just the invention of the telephone, television, and the Internet, but the harnessing of water, such as Niagara, for hydroelectric power!

Mother Shipton's own powers became so celebrated, they were recorded in the diaries of Samuel Pepys, the Member of Parliament made famous for his diary. Pepys recorded that, while surveying the damage done to London by the Great Fire of 1666 in the company of the Royal Family, he heard them discussing Mother Shipton's prophecy of the event, with no doubt some accompanying verse such as "A Great Fire will London's walls contain, magicked out of Pudding Lane," or some such rhyme (to be clear, I made this last verse up).

Indeed, on the question of fake news on Mother Shipton, her most famous prediction foretold the end of the world. "The world to an end shall come, in eighteen-hundred-and-eighty-one" appears to have been made up for profit, a process that had undoubtedly occurred to Mother Shipton in her day!

PREDICTIVE SCIENCE

Some degree of prophecy is also common to science. Science is a recipe for doing things. It revolves around its practical application, how things are done, and how outcomes can be predicted from practice. If you think about it, science tells us how to do certain tasks, should we need to do them, and what will happen when you do so. Clearly, there's great power in this philosophy. From knowing this philosophy of matter in motion, this account of nature from below rather than above, science tries to own the fate and power of change by getting to know nature's ground rules.

We talked earlier in the book about the way in which science fiction is an imaginative device for doing a kind of theoretical science: the exploration of imagined worlds. How scientists build models of hypothetical worlds, and then test their theories, and how science-fiction writers and moviemakers also explore hypothetical worlds of the future with more scope and fewer boundaries but using that spirit of "what if?" common to both science and science fiction. As science fiction is obsessed with the future, it's hardly surprising to find that prediction and precognition (literally *seeing* into the future) is a very popular motif. Perhaps Phillip K. Dick's 1956 short story *The Minority Report* is best known in this regard. While no one is exactly calling Dick the "Mother Shipton of sci-fi," his three "precog" mutants, who are able to divine precrime, are famously featured in the Steven Spielberg 2002 movie adaptation of *The Minority Report*. The "precogs" have abilities by which they can see up to two weeks into the future. They're strapped into machines, in which they nonsensically babble, while a computer listens and converts their gibberish into predictions of the future.

It may seem like science fiction, but a new field of science does something similar. Cliodynamics (named after *Clio*, the Greek Muse of history) enables scholars to look critically at history in the hopes of predicting the future. The technique foresees a wave of widespread violence around 2020, including riots and terrorism. This mushrooming field of cliodynamics uses the so-called dark archives, data banks from the distant past, to extrapolate based on historical trends.

Scholars in this new field have found patterns of social unrest. Analysis clearly shows 100-year waves of instability. Superimposed on each wave is an additional 50-year cycle of widespread political violence. The cycles of violence have social inequality at their root. Discontent builds up over a period of time until the pressure is violently released. Scholars have mapped the way that social inequality creeps up over the decades, so much so that a breaking point is reached. A little late, reforms are finally made, but over time those reforms are reversed, and society lurches back to a state of heightening social inequality. So much for the dark archives. One wonders what the cliodynamic scholars would make of this final rhyming prophecy from Mother Shipton; "And wives shall fondle cats and dogs, and men live much the same as hogs."

LORD OF LIGHT:

HOW DOES SCIENCE FICTION CHEAT DEATH?

"I believe in my own obsessions, in the beauty of the car crash, in the peace of the submerged forest, in the excitements of the deserted holiday beach, in the elegance of automobile grave-yards, in the mystery of multi-story car parks, in the poetry of abandoned hotels. . . . I believe in the power of the imagination to remake the world, to release the truth within us, to hold back the night, to transcend death, to charm motorways, to ingratiate ourselves with birds, to enlist the confidences of madmen."

—J.G. Ballard, poem "What I Believe" in
French magazine *Science Fiction* (1984)

"The [Sorcerer's] Stone was really not such a wonderful thing. As much money and life as you could want! The two things most human beings would choose above all—the trouble is, humans do have a knack of choosing precisely those things that are worst for them."

—Albus Dumbledore in J.K. Rowling's *Harry Potter
and the Sorcerer's Stone* (1997)

"I would love to believe that when I die I will live again, that some thinking, feeling, remembering part of me will continue. But as much as I want to believe that, and despite the ancient and worldwide cultural traditions that assert an afterlife, I know of nothing to suggest that it is more than wishful thinking. . . . The world is so exquisite, with so much love and moral depth, that there is no reason to deceive ourselves with pretty stories for which there's little good evidence. Far better, it seems to me, in our

vulnerability, is to look Death in the eye and to be grateful every day for the brief but magnificent opportunity that life provides."
—Carl Sagan, *Billions and Billions: Thoughts on Life and Death at the Brink of the Millennium* (1997)

"I would not live forever, because we should not live forever, because if we were supposed to live forever, then we would live forever, but we cannot live forever, which is why I would not live forever."
—Miss Alabama in the *Miss USA* contest (1994)

THE MORTAL IMMORTAL

Eloquent beauty pageant contestants notwithstanding, the search for immortality has been one of science fiction's most persistent obsessions. Immortals can be found in fairy tales and myths, of course, such as the wizard; Merlin; the Greek hero of the Trojan War, Achilles; and the Wandering Jew, a shoemaker who taunted Jesus on his way to the crucifixion and was cursed to "go on forever" until the second coming of Christ. Of late, however, the obsession with immortality has looked to science to cheat death.

Mary Shelley, ever the founding mistress of fictional ideas, features an undying hero in her 1833 story, *The Mortal Immortal*. Turning to science rather than magic, her hero's fate is to watch those around him wither and perish while he alone abides. He has effectively become a time traveler, fated to forever move forward. Shelley's story uses early thoughts on biology to achieve immortality, using treatments and medicines to artificially extend life span.

Indeed, since Shelley, science fiction has used many ideas and designs for eternal life. They have included eugenics, genetic engineering, the use of artificial organs, and even becoming a cyborg. This last option is typified by the Cybermen in *Doctor Who* who gained immortality by replacing their body parts with cybernetic ones but lost their humanity in the process.

Darwinism is hard to reconcile with immortality. Death is in the very detail of nature's plan. The fossil record has long promised that all species perish, and so in the Darwinian scheme of things, species are not fixed or everlasting. There is no difference between humans and the billions of extinct creatures from the dustbin of geological history, so how could humans possibly cheat history and go on to a life beyond nature?

CYBERNETICS OR GENETICS?

Aubrey de Grey is a leading biogerontologist. Biogerontology is a field that attempts to answer the question of why and how we age, and de Grey is confident he has the answer. This Cambridge University academic is so sure, in fact, that he has established the Methuselah Mouse Prize, a monetary award that goes to any researcher who can significantly extend the life of a mouse. De Grey sees the Prize as the first step in biological immortality. Mice first, men later.

Some writers wager on cybernetics fixing the aging problem. Of late, this has centered around the idea of computer tech that would allow us to upload our personalities to live forever. British writer Olaf Stapledon was one of the first to explore this idea in his 1930 novel, *Last and First Men*. In the book, humans come close to the immortal by developing human-like brains grown into a machine. *Lords of Light*, written by American writer Roger Zelazny in 1967, is a story about crew members of a space vessel who achieve immortality by uploading their consciousness into new bodies. They do this so many times that they start to see themselves as gods. Leading robotics and cognitive science engineers, influenced by science fiction, refer to this kind of new consciousness as an infomorph—a type of artificial intelligence best explored in the work of Australian science fiction writer Greg Egan. While scientists acknowledge that current technology isn't fit to tackle the infomorph just yet, some point to the exponential development of computing to suggest that the infomorph may soon be on the horizon.

If the infomorph doesn't give us everlasting life, maybe genetics will. Taking the idea that our genes codify who we are, an idea made popular by evolutionary biologist Richard Dawkins in his 1976 book, *The Selfish Gene*, the consequence may be that cloning becomes a method of achieving immortality. The science-fiction movies that explore this topic are suddenly cast into a new light—can identical genetic material be made to last forever?

You can always bet on quantum physics guys to throw a weird wrench in the works—i.e., "quantum immortality." According to some specialists in quantum technology, the many-worlds interpretation of quantum theory means that every action occurs (or doesn't) somewhere in an infinite set of universes. So, there exists a path through your life in which you never die, every decision you ever made leads to your continued life. I do hope, dear reader, that you are as unsatisfied with this option as I am writing it. The chance of this being true, let alone happening, is so phenomenally low, it's hardly worth me writing it. In fact, in most universes I probably haven't written this. It would be even weirder if, the next time you pick up this book to read this particular section, it's no longer here.

APOCALYPSE NOW:

IN WHAT SIX WAYS DOES SCIENCE FICTION SEE THE END OF OUR WORLD?

FADE IN:
TV images. A high street. An abandoned car. On the road beside it, blood. And a little girl's red shoe.

BEYOND:
A UK city run riot. A powerful virus has locked its citizens into a state of murderous rage. The INFECTED. Hordes of them are on the rampage, killing and maiming with an insatiable appetite for violence.

—Rowan Joffé, Juan Carlos Fresnadillo, and Jesús Olmo, draft script of *28 Weeks Later* (2007)

"Needles in a heavenly haystack. There are more stars in the heavens than human beings on earth. Through telescopes, men of science constantly search the infinitesimal corners of our solar system seeking new discoveries, hoping to better understand the laws of the universe. Observatories dedicated to the study of astronomy often are set in high and remote places. But there is none more remote than Mount Kenna observatory in this part of South Africa. 'If our calculations prove to be correct this will be the most frightening discovery of all time. These two bodies have traveled almost a million miles in two weeks.'"

—Sydney Boehm, screenplay, *When Worlds Collide* (1951)

[Looking through Shaun's LPs for suitable records to throw at two approaching zombies]
Ed: "Purple Rain?"

Shaun: "No."

Ed: "Sign o' the Times?"

Shaun: "Definitely not."

Ed: "The Batman soundtrack?"

Shaun: "Throw it."

Ed: "Dire Straits?"

Shaun: "Throw it."

Ed: "Ooh, Stone Roses."

Shaun: "Um, No."

Ed: "Second Coming."

Shaun: "I like it!"

—Simon Pegg and Edgar Wright, screenplay,
Shaun of the Dead (2004)

THE END OF THE WORLD

The end of the world is coming. Just how close we are, and by what means it will happen, is anyone's guess. You can be sure, however, that popular and scientific concern about the coming apocalypse has been driven by science fiction. While it's true that religious depictions of catastrophe appear in many faiths, in modern times it's science fiction that has conjured a sustained exploration of our demise, and the way in which our pale blue dot might be extinguished, or at least the technological apes living on it. Here are six ways in which science fiction has envisioned the end:

1: Plague

During 1816, as Mary Shelley sat writing *Frankenstein*, a book that some say is the first work of science fiction in the modern age, the skies outside were darkened by volcanic ash during that so-called "year without a summer," following the eruption of Mount Tambora. Shelley's second work of science fiction directly follows the apocalyptic events that might have been set in motion by the Tambora explosion. *The Last Man* (1826) deals with the effects of a worldwide plague, which drifts across the face of the globe, leaving just a sole survivor.

The horror of Shelley's vision of a plague annihilating mankind was brought to life with the 1918–1919 Spanish Flu pandemic. An estimated 5 percent of the world's population died, around 100 million souls, making it one of the deadliest natural disasters in human history. Science fiction needed no further encouragement. The last century was dominated by movies such as *The Andromeda Strain* (1971), *28 Days Later* (2002), and its sequel, *28 Weeks Later* (2007). The portrayal of future pandemic apocalypses has influenced the World Health Organization to plan and coordinate responses to future outbreaks across the world.

2: War

H.G. Wells was profoundly influenced by Mary Shelley's work. His own vision of the explosive devastation caused by an atomic bomb became the defining focus of concern about the end of the world for most of the twentieth century. Visions of atomic apocalypse were inspired by the infamous detonations in Japan in 1945, and the ensuing Cold War and its test explosions. Movies such as the adaptation of Nevil Shute's relentlessly morbid *On the Beach* (1959) capture the postwar mood of impending doom. The story is set in Australia, the only continent left unscathed by a nuclear war. Here too the fallout will soon engulf the land, and as the action ends with a poignant closing shot, a banner flutters in the breeze to a soundtrack of *Waltzing Matilda*. It reads, "There is still time . . . brother."

3: Natural Crisis

Thrillers such as *The China Syndrome* (1979) and *Silkwood* (1983) explored cover-ups at nuclear power stations and spoke of the risks of future societies going totally nuclear. Once more fiction became fact when in 1986 the Chernobyl Disaster led to a plume of radioactive debris drifting across the Soviet Union and Europe. Tales about the demise of the world, as we know, are often based on natural rather than technological reasons. Notable here is John Christopher's 1956 posta-pocalyptic novel, *No Blade of Grass*, in which a new virus strain infects crops, causing a massive famine. These days, if we mess up this planet, at least we know there are other worlds out there we can try traveling to. Unless the aliens come here first . . .

4: Alien Invasion

Exactly how they know we're here is anyone's guess, as they must surely have come from somewhere deep in space. The alien invasion subgenre of science fiction, however, focuses instead on what the aliens want to do to us. These greedy extraterrestrials invade Earth either to slaughter and supersede us, or to enslave us, or to harvest us as food, or, for no apparent reason at all, to destroy the planet altogether (in *The Hitchhiker's Guide to the Galaxy*, Douglas Adams jokes that the invading Vogon Constructor Fleet destroys the earth to make way for a hyperspace bypass through our planet).

5: Cosmic Catastrophe

In his 1987 lesser-known short story, "The Star," H.G. Wells writes of a strange luminous object that erupts into the solar system and makes its way toward a fragile earth. As the object looms ever closer, havoc is wreaked upon the earth when gravitational tidal waves cause global devastation. Similar tales were told in movies throughout the next century. *When Worlds Collide* (1951) is a film about the coming destruction of the earth by a rogue star and so bears a close resemblance to Wells's story. In July 1994, humans witnessed the first-ever extraterrestrial collision when Comet Shoemaker-Levy 9 ploughed into planet Jupiter. The collision was no doubt the inspiration for the "paranoia" of the two 1998 movies *Deep Impact* and *Armageddon*, where Bruce Willis again saves the world in his now-familiar vest.

6: AI

In Steven Spielberg's movie *AI: Artificial Intelligence*, rising sea levels and global warming drastically reduce the world's population. However, two millennia later, humans have become extinct, and Manhattan is buried under ice. Humans have been superseded by an advanced silicon-based intelligence known as Mecha. In *The Matrix* (1999), the demise of humans comes a lot earlier in the twenty-first century, as intelligent machines wage that traditional Frankenstein-like science-fiction war against their creators. This cult movie became so popular that whenever someone was spotted on a pay phone, people assumed they were looking for an exit from the Matrix.

On a final surreal note is the wonderful movie *Donnie Darko*. Here, a disaffected teenager, played by Jake Gyllenhaal, is told by a giant rabbit named Frank that the world will end in twenty-eight days. The journey that unfolds involves societal failures, sex, and time travel, which just goes to show the world can end in as many ways as there are people.

MAN IN THE HIGH CASTLE:
WHAT DOES SCIENCE FICTION SAY ABOUT HISTORY?

"Science fiction writers write it. And it uses a very science fictional technique: change one thing and extrapolate from that."
—Harry Turtledove on alternate history,
interview at WorldCon (2001)

"Jonbar Point: term used for a crucial forking-place in time, whose manipulation can radically affect the future that follows. The name derives from Jack Williamson's *The Legion of Time* (1938), which deals with the potential future empires of Jonbar (good) and Gyronchi (bad). The former is named for the character John Barr: the fiercely contested jonbar point is the moment when as a small boy Barr picks up either a magnet, inspiring him to a life of science which ultimately brings Jonbar into existence, or a pebble, leading Barr to obscurity and the world to Gyronchi."
—John Clute and Peter Nicholls,
The Encyclopaedia of Science Fiction (1979)

"Sometimes it's possible, just barely possible, to imagine a version of this world different from the existing one, a world in which there is true justice, heroic honesty, a clear perception possessed by each individual about how to treat all the others. Sometimes I swear I could see it, glittering in the pavement, glowing between the words in a stranger's sentence, a green, impossible vision—the world as it was meant to be, like a mist around the world as it is."
—Ben H. Winters, *Underground Airlines* (2016)

"There was once only the dust particles in space, the hot hydrogen gases, nothing more, and it will come again. This is an interval, ein Augenblick. The cosmic process is hurrying on, crushing life back into the granite and methane; the wheel turns for all life. It is all temporary. And they—these madmen—respond to the granite, the dust, the longing of the inanimate; they want to aid Nature. And, he thought, I know why. They want to be the agents, not the victims, of history. They identify with God's power and believe they are godlike. That is their basic madness."
—Philip K. Dick, *The Man in the High Castle* (1962)

"Shoot the dictator and prevent the war? But the dictator is merely the tip of the whole festering boil of social pus from which dictators emerge; shoot one, and there'll be another one along in a minute. Shoot him too? Why not shoot everyone and invade Poland? In fifty years', thirty years', ten years' time the world will be very nearly back on its old course. History always has a great weight of inertia . . . almost always."
—Terry Pratchett, *Lords and Ladies* (1992)

WHAT IF . . .

What if history had happened so very differently? What tiny amount of tinkering in time would result in the most profoundly different present? What if grass had died? What if Neanderthals had not become extinct? What if Earth had a habitable moon instead of a dead lunar satellite, or a series of alien invasions had made the world well aware of life on other planets before the Renaissance? What if Christopher Columbus never sailed west? What if the British Empire had conquered China? What if there were no oil in the Middle East? What if the South had won the US Civil War? What if Hitler had successfully invaded Russia?

British novelist L.P. Hartley once said, "The past is a foreign country; they do things differently there." In the science-fiction genre of alternate history, however, things in the past *happened* differently, and the present *becomes* a foreign country, of sorts. Alternate history stories contain "what if" scenarios, like the list above, which revolve around crucial

"Jonbar" points in the past that could have had a different outcome to that recorded in actual history.

One early use of an alternate history scenario appears in Edward Gibbon's classic work *The Decline and Fall of the Roman Empire* (1776–1778). Gibbon speculates what might have happened had Muslims won the battle of Poitiers (it was a very close call) in France in 733 AD: "The Arabian fleet might have sailed without a naval combat into the mouth of the Thames. Perhaps the interpretation of the Koran would now be taught in the schools of Oxford, and her pulpits might demonstrate to a circumcised people the sanctity and truth of the revelation of Mahomet."

The Book of Mormon (1830) could also be seen as alternate history. According to the *Book*, Old World migrants came to the Americas in several waves, mainly Jews from the Levant, and inhabited the continent from about 2000 BC to 400 AD. Allegedly, these mysterious migrants built sumptuous cities large enough to support hundreds of thousands of warriors, and Native Americans are mostly descended from them. The *Book* is a variation on a commonly held belief in early nineteenth-century America—namely, that the Americas were settled by Old World immigrants whose established and advanced culture fell into decline. Conveniently, all evidence of this culture had mystifyingly vanished by the arrival of Europeans in the 1490s. Understandably, *The Book of Mormon* narrative is regarded as nonhistorical by a wide range of scholars. To quote just one single piece of real science on the matter, the Native American nation known as the Navajo has a gene marker inherited from the Chukchi, a tribe that still lives deep in Arctic Russia. Scholars believe the Chukchi were among those who migrated from Asia to America, many millennia earlier than 2000 BC.

ALTERNATE HISTORIES

For most of the twentieth century, modern cultural accounts of alternate illness have been associated with science fiction. Typically, such histories blend with time travel. Jumping from one history to its alternate, and awareness of the presence of one time line by the people in another, is a common theme of the genre. Indeed, crosstime alternate histories have become so closely related that it's almost impossible to separate them from the genre as a whole. Perhaps the most influential early science-fiction alternate-history tale is H.G. Wells's *Men Like Gods* (1923). The story features travelers transported to an alternate Earth that is almost a utopia and that has diverged from our own history several centuries before. As a result of Wells, time line-hopping tales became incredibly popular in early pulp fiction and through the twentieth century.

Another notable alternate history is *The Difference Engine* by William Gibson and Bruce Sterling. The novel presents a history in which Charles Babbage, the patron saint of the programmable computer, transforms the Victorian world with his inventions. London becomes a protocybercity as the result of the successful production of Babbage's analytical engine, an early computer that didn't quite make the grade in our own time line. Gibson and Sterling's counterfactual fiction imagines the mass production of card-driven computers that transform western society and fuel a turbo-charged industrial revolution. These different engines lead to a changed world, and the information revolution of the late twentieth century is played out against a steam-punk setting. Data dance between steam-sheathed turbines as every man, woman, and child is marked and counted: "Behind the glass loomed a vast hall of towering Engines—so many that at first Mallory thought that the walls must surely be lined with mirrors like some fancy ballroom. It was like some carnival deception, meant to trick the eye—the giant identical engines, clock-like constructions of intrinsically interlocking brass, big as railcars set on end, each on its foot-thick padded blocks. The whitewashed ceiling, thirty feet overhead, was alive with spinning pulley-belts, the lesser gears drawing power from tremendous spoked flywheels on socketed iron columns. White-coated clackers dwarfed by their machines, paced the spotless aisles."

Gibson and Sterling's alternate history is similar to our world's black tarmacadamed streets of Victorian London, and yet they conjure the great city into an even grimier smog-stained version of the metropolis. Though their take on those Dickensian days seems strikingly similar to our own time, a distance between the time lines is cleverly struck with their "clackers," rather than "hackers," and their mean London back alleys that invest their world with considerable swagger. These aren't dull Victorians, they're Victorians with attitude. But, behind the science-fictional swagger,

Gibson and Sterling are making a serious point. They are pushing the origin of steam-punk back into the Victorian past. They're relocating the matrix back into those days of Dickens, when change thrummed along electric wires and the ubiquity of the modern computer found its antecedents in the pioneering science of the Victorian era.

THE NAZIS: A WARNING FROM ALTERNATE HISTORY

Fiction's most common change point, at least in the English language, is "What if Germany (or Japan) had won World War II?" (the next most popular being "What if the South had won the Civil War?" and "What if the Spanish Armada had not been defeated?"). Taking the Nazi victory in the Second World War as an example, we can see the different ways in which counterfactual histories give writers a chance to make social and political speculations on history. Robert Harris's *Fatherland* (1992) is a single past that has a sense of history enabled through the creation of a plausible alternative: Nazi Germany successfully invades Russia in 1942. Learning that Britain has broken the Enigma code, the Nazis play it safe and broker for peace with the west. Through the magic of propaganda, Hitler is revered twenty years later as a beloved leader. It's an alternate history, of course, but Harris was drawing a parallel with real history: the world of *Fatherland* is Stalin's Russia with the names changed. In Harris's alternate history, however, the characters are unaware of any other time line.

Philip K. Dick's classic *The Man in the High Castle* (1962), on the other hand, is a more complex Nazi counterfactual. Dick's book, rather than portraying a single time line, has some of its characters aware that other time lines exist, and that they may be in the wrong history. Dick's story poses questions about the authenticity of reality and the truth of history itself. *The Man in the High Castle* helped develop alternate history fiction as a serious genre.

A troubled man, Dick spent most of his years battling with mental health, a theme he was to visit in much of his work. Alongside the sequence of broken relationships that littered his life, the realities in Dick's fiction have a fractured feeling, with autobiographical elements featuring strongly in his texts. Set in 1962 (the year the novel was published), Dick's *The Man in the High Castle* portrays a divided United States, partitioned between the victorious Axis Powers of Nazi Germany and the Empire of Japan. Interweaving several story lines, the main thrust of the novel focuses on espionage and attempts by the Japanese to discover a Nazi plot to launch a nuclear attack on their home islands.

It is within the detail of the domestic lives of Americans, however, that Dick's novel contemplates the question of history. Central to this contemplation is Frank Frink, who manufactures and distributes fake Colt pistols as authentic artifacts of the American Civil War. The Japanese occupiers are keen clients, desperate for some sense of real local history. When one of Frank's fake guns is used by a client to successfully defend himself from Nazi agents, it calls into question the notion of what is real. Ray Calvin, another character involved in the fake-memorabilia scam, makes this point to a girl. He gives her two very similar-looking cigarette lighters, one of which is worth "maybe forty or fifty thousand dollars on the collector's market." Why? Because of the historicity. She says, "What is historicity?" To which he responds, "When a thing has history in it. Listen. One of those two Zippo lighters was in Franklin D. Roosevelt's pocket when he was assassinated. And one wasn't. One has historicity, a hell of a lot of it. As much as any object ever had. And one has nothing.... You can't tell which is which. There's no 'mystical plasmic presence,' no 'aura' around it." The notion of what is real and what is fake extends beyond mere artifacts and out into the realities that the characters inhabit. The book's climax of this dilemma is the pivotal section of *The Man in the High Castle*, when Frink encounters a best-selling underground novel entitled *The Grasshopper Lies Heavy*, in which it becomes clear that other potential realities exist, including one where the allies won the war.

As a footnote, it's worth noting that the most popular counterfactual in French is "What if Napoleon had not been defeated?" This is unusual, as alternate histories always seem to throw up realities far worse than ours, which makes you wonder about the political intentions of authors who dream about a Nazi victory, or a triumphant South in the American Civil War. This favorite French alternate history results in a better world than the one we have. Why? One commentator remarked, "We look at this as the best of all possible worlds, but the French know it isn't, because most people speak English."

PART III
MACHINE

If historians in the future were asked to characterize this age of ours by any single expression, they'd probably call it the early Machine Age.

It's more than three million years since the start of the Stone Age, but only three hundred years or so since the invention of the early steam engine. Sometimes known as the *philosophical* engine, as it was based on Newton's system of the world, the steam engine reshaped the planet. Through the force of fire, the engine allowed our first uncertain steps into an industrial way of life. The engine drove locomotives along their metal tracks and propelled steamships across the Atlantic. Machines enabled the building of better bridges and roads, triggered the telegraphs that ticked intelligence from station to station, and lit up the iron foundries and coal mines, which powered this Industrial Revolution.

As all the machinery began to mesh, and science encroached upon every aspect of life, progress and technology seemed inseparable. For every factual gadget, science fiction spawned a thousand visions. Early optimism about the machine evaporated as the mood of the age changed. Science fiction was deeply divided into trends of light and darkness, as writers began to come to terms with the double-edged sword of technology and change. This mood was epitomized by Mary Shelley's *Frankenstein*, a seminal voice on the strange fruits of science in the scarred landscapes and dark satanic mills of industrial England. *Frankenstein* called for vigilance in the practice of science. It warned of the primal urges of power and control in all creations of technology. The book became a potent metaphor of the powerlessness of the inventor. Ever since *Frankenstein*, codified in such artifacts as the hydrogen bomb and human genetics, it has proved difficult to limit the social fallout of science.

Mary Shelley's field of interest was the conflict between the human and the nonhuman. It is unsurprising that she was part of the Romantic Movement. Most mainstream fiction since the Renaissance has been unconcerned with the Universe revealed by science. Poetry has little to do with the laws of physics, they would argue. For the Romantics, however, and for science fiction, the dialogue between the human and the nonhuman is the key concern. Way back in 1798 in his *Lyrical Ballads*, English Romantic poet William Wordsworth hints at the science fiction of the future. "If the labors of men of science should ever create any material revolution . . . in our condition . . . the poet will sleep then no more than at present, but he will be ready to follow the steps of the man of science, not only in those general indirect effects, but he will be at his side, carrying sensation into the midst of the objects of the science itself." Trying to best express "the taste, the feel, the human meaning of scientific discoveries," as Wordsworth put it, is exactly how science fiction works. It has presented a mode of thinking concerned with the reducible gap between the new worlds uncovered by science and the fantastic, strange worlds of the imagination.

From the very start of the Machine Age, science fiction concerned itself with technology. How do we create devices without sacrificing some of what it means to be human? When do humans and machines achieve a symbiosis so that they become a new form of life? Is technology neutral, or can some machines truly be described as evil? Today, we are still wary of machines and machine intelligence, perhaps for good reason. There is little doubt, however, that one of the main reasons for our skepticism about machines is the way they are portrayed in science fiction. The gadgets we see fluttering across the silver screen are rarely designed to tuck us safely into our beds at night. Instead, a legion of replicants, droids, and AIs advance across our darkest imaginings. They seem hell-bent on disemboweling humans and using their entrails as a hat, if machines actually *wore* hats, of course. So, when we jack into the many virtual worlds depicted in science fiction, we meet psychotic machines, single-minded about mechanical mayhem and determined to take over the world.

One of science fiction's most famous machine inventions is the robot. Yet again, the mad inventor manufactures a machine, in this case a humanoid automaton, only to see it rise up against its master and deliver the clearest of messages: the creation of new technology has a dark side. Science fiction sees light in the machine, as well as darkness. Its past is littered with creative attempts to imagine a future in which machines are our friends, utopian dreams of gleaming metal spires housing legions of labor-saving droids, toiling industrially to serve our every whim. Yet, it is the dark side of the machine that, more often than not, wins out in fictional tales of technology. Perhaps we have become too mistrustful of the machine, too reliant. In this world where we depend on them for communication, transport, medicine, and almost every other walk of life, maybe we realize how much we would miss that reliance.

No other machine in science-fiction history has been as enthusiastically embraced as the spaceship. Perhaps more than any other device, the spaceship reminds us that humans are, at heart, inventors and explorers. In a perfect evocation of the relationship between science and science fiction, one of the most influential science-fiction franchises of all time, *Star Trek*, was based on the words of Dr. James Killian, science adviser to President Eisenhower, who wrote in the proposal for a national space program that "It is useful to distinguish among four factors which give importance, urgency, and inevitability to the advancement of space technology. The first of these factors is the compelling urge of man to explore and to discover, the thrust of curiosity that leads men to try to go where no one has gone before." Acknowledging their debt to *Star Trek* for helping popularize the human journey into space, NASA in 1976 bowed to pressure from fans of the show in the form of a write-in campaign and changed the name of its first space shuttle from *Constitution* to *Enterprise*.

Within the machine chapters of this book, you will find examples of some of the ideas, principles, and machine technologies that have crossed from fiction to fact and back again. To claim that a single scientist or author is the sole inventor of a technology is to ignore and dismiss the contributions of armies of scientists and writers who helped make the initial spark become a reality. Mashed up with "big picture" concepts such as rockets and robots, and the atom bomb and the machine state of *1984*, you'll find the world of entertainment represented by cyberspace and virtual reality, as well as the prospect of finally driving a flying car.

All these and more have contributed to the creation of our contemporary culture. From the way we almost instantly communicate with the world, to the prospect of living a life without having to talk to anyone at all, science fiction has been busy influencing science and culture and, as you shall see, a future in which the machine abides, if not masters.

BATTLE OF NEW YORK:
HOW DID SCIENCE FICTION INVENT THE NUKE?

"Nothing could have been more obvious to the people of the early twentieth century than the rapidity with which war was becoming impossible. And as certainly they did not see it. They did not see it until the atomic bombs burst in their fumbling hands."

—H.G. Wells, *The World Set Free* (1914)

"Despite the vision and farseeing wisdom of our wartime heads of state, the physicists have felt the peculiarly intimate responsibility for suggesting, for supporting, and in the end, in large measure, for achieving the realization of atomic weapons. Nor can we forget that these weapons, as they were in fact used, dramatized so mercilessly the inhumanity and evil of modern war. In some sort of crude sense which no vulgarity, no humor, no overstatement can quite extinguish, the physicists have known sin; and this is a knowledge which they cannot lose."

—J. Robert Oppenheimer, *Physics in the Contemporary World*, from Arthur D. Little Memorial Lecture at MIT (November 25, 1947)

"When you see something that is technically sweet, you go ahead and do it and you argue about what to do about it only after you have had your technical success. That is the way it was with the atomic bomb."

—J. Robert Oppenheimer, testifying in 1954, discussing whether America should develop the "super" hydrogen bombs with vastly higher explosive power; from *The Oppenheimer Hearing Transcripts*

"The 20th century was a test bed for big ideas—fascism, communism, the atomic bomb . . . 'Greater than the tread of mighty armies is an idea whose time has come,' said Victor Hugo. In either case, run."

—P.J. O'Rourke, "Let's Cool It with the Big Ideas"
from *The Atlantic* (2012)

BATTLES OF NEW YORK

There have been a number of battles of New York. In 1776 and 1777, there was a run of battles for control of New York City and New Jersey in the American Revolutionary War. British commander Sir William Howe first defeated George Washington's Continental Army but then overextended his reach into New Jersey. The other major Battle of New York occurred in the Marvel Universe, when Loki looked like he'd take the city only to overextend himself in trying with his Chitauri army to take Manhattan against the Avengers. The striking difference between the fact and the fiction is the presence of a superweapon: the nuke, a weapon curiously first invented in science fiction, and not in science.

"And these atomic bombs which science burst upon the world that night were strange, even to the men who used them." So quoted H.G. Wells in his prophetic 1914 novel, *The World Set Free*. Wells's book was the first to christen the "atomic bomb," and it was a story that led to Hiroshima. By the turn of the twentieth century, it had become clear that some form of atomic energy was responsible for powering the sun and the stars. In 1899, American geologist Thomas Chrowder Chamberlin wrote in the journal *Science* that atoms were "seats of enormous energies" and that "the extraordinary conditions which reside in the center of the sun may . . . set free a portion of this energy."

In 1903, the great New Zealand-born nuclear physicist Ernest Rutherford and his coworker, Frederick Soddy, had been the first to calculate the vast amount of energy released in radioactive decay. Both were alive to the idea that this energy was potentially lethal. Indeed, Rutherford is alleged to have claimed that some clown in a lab might blow up the world unawares. In a 1904 lecture quoted in his book *Atomic Transmutation*, Soddy reasoned that "The man who put his hand on the lever by which a parsimonious Nature regulates so jealously the output of this store of energy would possess a weapon by which he could destroy the Earth if he chose." While Soddy trusted nature to "guard her secret," H.G. Wells took a different approach. After all, Wells had tales to tell.

Wells's timetable for the development of nuclear capability is unnervingly far-sighted, if not a little conservative in his dates. In Wells's *The World Set Free*, it wasn't until the 1950s that a scientist uncovers atomic energy and realizes there's no going back from this momentous discovery. Nonetheless, the scientist feels "like an imbecile who has presented a box of loaded revolvers to a crèche." Wells's book also foresaw a world war in 1956, with an alliance of France, England, and America against Germany and Austria.

There had been superweapon fiction before Wells, but it had been prone to cliché, and the naïve notion that the obsessive mind of a single genius could change the course of history. Human problems could be solved by the technofix of a scientific miracle, these stories suggested. Wells, however, was wise enough to know that this level of technical advance does not come from the know-nothing notion of genius. It comes from the dialectic between nations and their productive forces, and Wells correctly predicted the emergence of the military-industrial complex to build the bomb.

ATOM BOMB

Wells also predicted a holocaust. In *The World Set Free*, some of the world's principal cities are obliterated by small atomic bombs dropped from airplanes. Wells wasn't guessing, either, for the weapons Wells portrays are truly nuclear. They use Einstein's equivalence of matter converted into fiery and explosive energy, all triggered by a chain reaction.

Wells's far-sighted novel was Szilárd's guiding light. After reading *The World Set Free* in 1932, Szilárd became the first scientist to seriously examine the nuclear physics behind the fiction. Szilárd became incensed after reading an article by Rutherford in the London *Times*, which rejected the idea of using atomic energy for practical purposes. His fury, fused with his legendary quick wit, enabled Szilárd to eureka the details of the very nuclear chain reaction that Rutherford denied. Legend has it that Szilárd was so smart he'd worked out the reaction while waiting for traffic lights to change on Southampton Row in Bloomsbury, London. One year later, Szilárd filed for a patent on the concept. He became the driving force behind the Manhattan Project. It was his idea, along with Einstein, to send a letter in August 1939 to Franklin D. Roosevelt, pointing out the possibility of atomic weapons. The two brilliant and influential Jewish scientists feared the irresistible rise of a Nazi Bomb.

Soon after, the Manhattan Project was masterminded. It would ultimately engage around 130,000 employees and cost a total of $2 billion (over $20 billion in today's figures). The project outcomes? Detonation of three atomic weapons in 1945: the Trinity test detonation in July in New Mexico; a uranium bomb, "Little Boy," detonated on August 6 over Hiroshima; and a plutonium bomb, "Fat Man," discharged on August 9 over Nagasaki.

Wells's fiction became factual terror over Japan. The bomb burst over Hiroshima as the city's 320,000 inhabitants were waking up. Thousands were slain in a second. Vaporized by light and energy: heat death. Szilárd had hoped that President Truman would merely "demonstrate" the bomb. Not use it against cities, as in Wells's *The World Set Free*. As the war raged on, however, scientists lost the power over their research.

The Manhattan Project's lead scientist, Robert Oppenheimer, mulled over the "atomic bomb" first realized in fictions and spoke for a generation of scientists when he said, "In some sort of crude sense which no vulgarity, no humor, no overstatement can quite extinguish, the physicists have known sin; and this is a knowledge which they cannot lose."

In the Atomic Age that followed 1945, the world lived with a mood of apocalyptic terror. Millions of people responded anxiously to the catastrophic visions of science fiction, and to the reality of the apocalyptic threat of the bomb that fiction had been so instrumental in developing. At this focal moment in history, many felt that science fiction was alone in its ability to project ways out of this predicament. It became the means by which a mass audience was confronted with the possibility of holocaust and mutually assured destruction. No other genre of literature came close.

IS GEORGE ORWELL'S *1984* BECOMING A REALITY?

"War is peace. Freedom is slavery. Ignorance is strength."
—George Orwell, *1984* (1948)

"It was possible, no doubt, to imagine a society in which wealth, in the sense of personal possessions and luxuries, should be evenly distributed, while power remained in the hands of a small privileged caste. But in practice such a society could not long remain stable. For if leisure and security were enjoyed by all alike, the great mass of human beings who are normally stupefied by poverty would become literate and would learn to think for themselves; and when once they had done this, they would sooner or later realize that the privileged minority had no function, and they would sweep it away. In the long run, a hierarchical society was only possible on a basis of poverty and ignorance."
—George Orwell, *1984* (1948)

"Don't you see that the whole aim of Newspeak is to narrow the range of thought? Every year fewer and fewer words, and the range of consciousness always a little smaller. Even now, of course, there's no reason or excuse for committing thought-crime. It's merely a question of self-discipline, reality-control. But in the end, there won't be any need even for that . . . by the year 2050, at the very latest, not a single human being will be alive who could understand such a conversation as we are having now."
—George Orwell, *1984* (1948)

"Power is in tearing human minds to pieces and putting them together again in new shapes of your own choosing."
—George Orwell, *1984* (1948)

"Something we don't have [is] a picture of the future. We have a system we are dissatisfied with. We know it's somewhat odd, we know is cracking in ways and is sometimes quite fake, especially with our politicians. But we have no other picture of the future, that's the problem, and the engineering system of the internet does not supply it. It's beautiful in other ways and it's great at organizing people, but we need a picture of the future somehow."

—Adam Curtis, on *HyperNormalisation* (2016)

1984

Big Brother is as famous as Frankenstein. Newspeak is as forked-tongue as fake news. And the legions of fictional Thought Police who read your mind are early echoes of the twenty-first century phenomenon of Hate Crime. From the telescreen to Room 101, today's world is strikingly similar to George Orwell's *1984*. We live in Big Brother's realm of fake news, political spin, and euphemism. War is conflict. Civilian casualties are described as "collateral damage." Lying politicians are now simply "misspeaking" or being "economical with the truth."

Few science-fiction novels have been as prophetic as Orwell's haunting specter of big government gone mad with lust for power. No novel written in the twentieth century has captured the popular imagination like *1984*. The very title of this classic science-fiction dystopia became a cultural catchphrase. The word Orwellian still ominously speaks of matters hostile to a free society.

The key to Orwell's story was his belief in a "catastrophic" future. It is a future of boundless despair. The book confronts the prospect of three totalitarian power blocks bringing history to a standstill. Big Brother is unassailable: "If you want a picture of the future, imagine a boot stamping on a human face—forever." In his essay *You and the Atomic Bomb* published in October 1945, just months after Hiroshima and Nagasaki, Orwell had written with exceptional insight about the age of atomic weaponry that lay ahead. It was clear that he was preparing the heart of darkness within *1984*: "We have before us the prospect of two or three monstrous super-states, each possessed of a weapon by which millions of people can be wiped out in a few seconds, dividing the world between them. It has been rather hastily assumed that that means bigger and bloodier wars, and perhaps an actual end to the machine civilization."

Orwell's vision was one of superpowers colluding and in tacit agreement never to use the bomb. Each superpower is apparently the archenemy of the other two. The truth is different. In fact, the power blocs "prop one another up like three sheaves of corn." Like the latter-day "War on Terror," continuous and limited wars allow each super-state to maintain hysteria within their borders. By the time *1984* was in print, the "Cold War" was a reality; indeed, Orwell actually invented the phrase in his essay *You and the Atomic Bomb*. The Iron Curtain had fallen, and Orwell's account of the politics of power blocs was staggeringly perceptive.

THE PARADISE OF LITTLE FAT MEN

Orwell's *1984* became a mirror of the fears and frustrations of the individual caught up in a complex, overly rationalized society. It was a prophecy of a totalitarian future based on "not any particular country, but the implied aims of industrial civilization." Orwell was troubled by the devastating effects of science and technology: "Barring wars and unforeseen disasters, the future is envisaged as an ever more rapid march of mechanical progress; machines to save work, machines to save thought, machines to save pain, hygiene, efficiency, organization . . . until finally you land up in the by now familiar Wellsian Utopia, aptly caricatured by Huxley in *Brave New World*, the paradise of little fat men."

In his 1932 novel, *Brave New World*, Aldous Huxley had imagined a technology that titillated. Orwell foresaw the technology of control. The insidious nature of *1984*'s culture of surveillance stems from its telescreens and Thought Police. In Orwell's marvelous words, "The Beehive State is upon us, the individual will be stamped out of existence; the future is with the holiday camp, the doodlebug, and the secret police." The Party of Big Brother fakes news, rationalizes language, and perverts history. Time is tampered with, dates of events forgotten or unascertainable. The science of information is used to maintain political control, underlining Orwell's point that "The really frightening thing about totalitarianism is not that it commits 'atrocities' but that it attacks the concept of objective truth: it claims to control the past as well as the future."

In *1984*, science's mastery of the machine is so complete that utopia is possible, but poverty and inequality are maintained as a means of sadistic control. The visual medium of monitoring in the two-way telescreen is a brilliant evocation of the all-seeing eye. In Orwell's book it is politicized into a technological nightmare. As Winston Smith dutifully follows the daily exercises on the telescreen, he is at the same time observed by it.

NOVEL SUPERWEAPON

1984 is a flawed masterpiece. Orwell had identified a new dark age. He saw the necessity for a social side to technological progress and to "reinstate the belief in human brotherhood." The Cold War had, however, already created the need for an ideological superweapon. Orwell's book was used with little regard to the author's intention. Many critics suggest that this was made far easier by the book's unrelenting portrayal of defeat. Neither Orwell's protagonist Winston Smith nor the "proles" have a ghost of a chance against the mechanical horror. The novel reinforces passivity rather than undermines it.

As a result, Orwell's intended warning had become a piercing shriek announcing the advent of the Black Millennium, the Millennium of damnation. This shriek, distorted by the mass media, frightened millions. The novel's ambitious British television adaptation alone, broadcast on BBC Television in the winter of 1954, was watched by an audience of over nine million viewers. The production was hugely controversial. Questions were asked in the British Parliament. Many viewers complained about the supposed subversive nature and horrific content. Rather than advance understanding, Orwell's book had become a prominent item in the propaganda of Hate Week. Millions continued to see the conflict of East and West in terms of black and white. Shaken by the public response to *1984*, Orwell issued a disclaimer from his hospital bed shortly before his death: "The danger lies in the structure imposed on socialist and on liberal capitalist communities by the need to prepare for total war . . . and by the new weapons, of which of course the atomic bomb is the most powerful." For Orwell, there was also danger in the acceptance of a totalitarian outlook by intellectuals of all colors: "The moral to be drawn from this dangerous nightmare situation is a simple one: don't let it happen. It depends on you."

The system of machine surveillance in Orwell's fiction swiftly became fact. *1984* became the standard text for describing the militarization of life. In 1954, US historian of science Lewis Mumford declared the world of Big Brother to be "already uncomfortably clear." American social scientist William H. Whyte cited Orwell's influence in his 1956 *The Organisation Man*, a best-selling study of corporate dictatorships such as General Electric and Ford. Sociologist David Riesman credited the popularity of dystopias to Orwell and the bomb: "When governments have power to exterminate the globe, it is not surprising that anti-Utopian novels, like *1984*, are popular, while utopian political thought . . . nearly disappears."

MERIDIAN OF THE MACHINE

In this twenty-first century, Orwell's machine vision may meet its meridian in the real world. According to Google engineer François Chollet, the Internet has become a kind of telescreen. Despite its advantages, the Internet is a totalitarian panopticon, a prison that uses the consumption of digital information as a psychological power vector so that people can be controlled from the center. Chollet says the online world is being shaped by two long-term trends. First, our lives are increasingly dematerialized. Both at home and at work, we constantly consume and generate data. Second, AI is getting ever smarter. These two trends mean that hidden social media algorithms control, to an ever-increasing extent, which articles we read, whom we contact, whose opinions we listen to, and whose feedback we see. In effect, control of our political beliefs and worldview. This curation of the data we consume gives the systems in charge increasingly more power over our lives and who we are. As the greater fraction of our lives creeps steadily into the digital realm, we become vulnerable to that which rules it.

Machine control is made all the easier by the fact that the human mind is highly vulnerable to social manipulation. Such techniques are nothing new, as Orwell and Huxley were able to predict them in their works of science fiction. As the human mind is a static, vulnerable system, however, and AI algorithms become ever smarter, this system could eventually simultaneously have a complete view of everything we do and believe, and complete control of the information we consume. In the tweeted words of François Chollet, "Importantly, mass population control—in particular political control—arising from placing AI algorithms in charge of our information diet does not necessarily require very advanced AI. You don't need self-aware, super-intelligent AI for this to be a dire threat."

Then why hasn't *1984* already happened? Simply because information systems haven't yet achieved the level of technological control in Orwell's vision—but that is about to change. AI is moving fast, and progress is feeding through into the deployment of more finessed targeting algorithms and social media bots. Deep learning has just begun to make its way into ad networks and newsfeeds. Social media giants, such as Facebook, have been investing huge amounts in AI research and development, with the explicit aim of dominating the field.

As François Chollet declared on Twitter in March 2018, "Who knows what will be next. . . . We're looking at a powerful entity that builds fine-grained psychological profiles of over two billion humans, that runs large-scale behavior manipulation experiments, and that aims at developing the best AI technology the world has ever seen. Personally, it really scares me. If you work in AI, please don't help them. Don't play their game. Don't participate in their research ecosystem. Please show some conscience . . ."

WHEN WILL WE BE ABLE TO BOAST BLADE RUNNER-LIKE FLYING CARS?

"My visions of the future are always pretty much standard issue. The rich get richer, the poor get poorer, and there are flying cars."

—Joss Whedon in David Lavery's *Joss Whedon, A Creature Portrait* (2014)

"Flying cars are not a very efficient way to move things from one point to another."

—Bill Gates, *Wired* (2013)

"We could definitely make a flying car—but that's not the hard part. The hard part is, how do you make a flying car that's super safe and quiet? Because if it's a howler, you're going to make people very unhappy."

—Elon Musk, *NBC News* (2014)

"The new millennium sucks! What a disappointment! What's the difference between the old millennium and the new millennium? Nothing! It's the same load of crap with a "2" in the front. When I was young, I said, 'I'm gonna live through a change! A massive change! Things are gonna be different! Things are gonna be great!' Screwed again! No flying cars! No flying cars!"

—Lewis Black, *Comedy Central Presents* (2000)

THE DREAM

Some people are never happy, are they? The future we were promised hasn't really arrived, they say. According to them, we're now meant to be living in a world of silver flame-retardant jump suits, ray guns, and X-ray specs. No doubt we're also meant to be invisible and immortal.

Some go even further. They believe the time has come to hold science fiction to task. This may be a world of cool technologies, they go on, but, dude, where's my flying car?

It's certainly true that few machines symbolize the future like the flying car. More than just a vision, flying cars are a kind of totem, a sign of things to come, an emblem of a bright, tech-laden future full of freedom and opportunity. That's probably why their lack of appearance so far has been felt so strongly by some. Jerry Seinfeld and George Costanza lamented the lack of such cars in a 1998 episode of *Seinfeld:* "It's like we're living in the 50s here!"

The earliest story equipped with a flying car was probably Jules Verne's *Master of the World* (1904). Verne's car not only flew, it also doubled as a boat or submarine. Flying cars became so ubiquitous that they eventually made popular appearances in futuristic children's TV programs such as the 1960s animation *The Jetsons* and later in a CGI reboot of Gerry Anderson's 1960s series *Thunderbirds Are Go*, even a Rolls Royce was capable of seamless flight.

In cinema too, the flying car has been seen "chittying" along for decades. Such visions run back to the late 1920s, when Fritz Lang's *Metropolis* showed ordered lanes of futuristic cars flying above the city. The cars looked a little like planes and bore more than just a passing resemblance to the traffic ambience glimpsed in the gleaming skyscapes on Coruscant decades later, so the fantasy of flying cars is with us still. They populate the cityscapes of Blade Runner's future-Los Angeles (1982), a visionary New York in *The Fifth Element* (1997), and the skies of a bit-magical Britain in the *Harry Potter* franchise.

THE REALITY

There can be no time more deserving than today for the introduction of the flying car, in our overpopulated world packed with traffic jams. Take China, for instance. September 2013 saw the world's longest traffic jam, more than 100 kilometers long, and lasting for weeks. The traffic problem is so common in China that some people have seen it as a business opportunity. Their motorbike business will weave its way between the gridlocked lanes and take you to the destination you were meant to be going to in the first place. Heck, they'll even provide someone to sit in your car for you until the jam is over (if it ever ends). Pizza deliveries to jammed cars are also very common. Pizza express, even if the traffic isn't.

This is clearly getting ridiculous. In a congested world such as ours, with a host of factors outside our control, no wonder people still dream of the freedom of the sky, made possible by the flying car. A 2010 MSNBC poll in the US found that 90 percent of the country would buy a flying car if given the chance.

It's not as if we've never tried building one. As early as 1928, a year after Fritz Lang's *Metropolis*, Henry Ford had realized the concept in an experimental single-seat aero plane he called a "sky flivver." According to an article in the London *Financial Times* in 2017, Ford had predicted in 1940, "Mark my words, a combination aero plane and motor car is coming. You may smile, but it will come." Yet, the first attempts with Ford's "sky flivver" were troubled; a pilot died in an early test flight. In 1956, cruise missile engineer Moulton Taylor unveiled the "Aerocar." Cruising up to 100mph, the little yellow car proved far more impractical than its fictional versions.

The flying car is technically doable, of course, but tricky in practice. One of the more recent real-life models is the futuristic "Skycar M400," invented by Paul Moller, who has been trying to perfect the flying car for more than fifty years. The Moller Skycar has a vertical, helicopter-like take-off and landing, smooth flight, and comfortable drive. It mostly transports four people, but single-seat and six-seat versions are also planned. It sounds "sci-fi" in its completeness.

So, what's the problem? Well, first, one of these beauties would cost you half a million dollars. Not only that, but every time a public demo has been announced, it has then been swiftly cancelled. It seems that fifty years on, the flying car is the best real-world example of "vaporware," a name the computer industry gives to a product that is announced to the public but never actually released. In this world of sustainability, fuel costs, and air traffic control, we're still some way off those mesmerizing skylines full of flivvers. So, dude, where's my flying car?

BREAKING NEWS

Mere days after I'd finished writing this entry, Chinese drone manufacturer Ehang released news of their Ehang 184 craft. In February 2018, the 184 flew forty journalists and officials on journeys of up to 15 kilometers in Guangzhou, southern China. The 184 is reported to have reached top speeds of around 80 mph. To fly the 184, passengers simply pinpoint their destination on a map, and the drone dutifully creates and executes a suitable flight plan. Ehang says the 184 is robust enough to weather thunderstorms and typhoons. Though, in the event of any problems, the craft's system can revert to remote pilot from a control center. So, dude, after all, HERE'S your flying car!

TOTAL RECALL:

WHEN WILL WE VACATION IN CYBERSPACE?

"Get ready for the ride of your life."
—*Total Recall* poster (1990)

"He awoke—and wanted Mars. The valleys, he thought. What would it be like to trudge among them? Great and greater yet: the dream grew as he became fully conscious, the dream and the yearning. He could almost feel the enveloping presence of the other world, which only government agents and high officials had seen. A clerk like himself? Not likely.
—Philip K. Dick, *We Can Remember It for You Wholesale* (1966)

"'So, you want to have gone to Mars. Very good. . . . You get tangible proof of your trip. . . . All the proof you'll need. Here; I'll show you.' He dug within a drawer of his impressive desk. 'Ticket stub.' Reaching into a manila folder, he produced a small square of embossed cardboard. 'It proves you went—and returned. Postcards.' He laid out four franked picture 3-D full-color postcards in a neatly-arranged row on the desk. 'Film. Shots you took of local sights on Mars with a rented moving camera.' Plus, the names of people you met, two hundred postcards worth of souvenirs, which will arrive—from Mars—within the following month. And passport, certificates listing the shots you received. And more . . . You'll know you went, all right . . . You won't remember us, won't remember me or ever having been here. It'll be a real trip in your mind; we guarantee that. A full two weeks of recall; every last piddling detail. Remember this: if at any time you doubt that you really took an extensive trip to Mars you can return here and get a full refund. You see?"
—Philip K. Dick, *We Can Remember It for You Wholesale* (1966)

"It's not quite telepathy, but a group of scientists have successfully eavesdropped on our inner thoughts for the first time. Using a newly designed algorithm, researchers were able to work out what people were saying in their heads based on brain activity. The idea behind this is not to give people Charles Xavier-style X-Men powers, but to eventually use such a system to help individuals who can't speak, for example due to paralysis, communicate with others. The work has been published in *Frontiers in Neuroengineering*."

—Paul Bence, *IFLScience!* (2014)

"He didn't really like travel, of course. He liked the idea of travel, and the memory of travel, but not travel itself."

—Julian Barnes, *Flaubert's Parrot* (1984)

WE CAN REMEMBER IT FOR YOU WHOLESALE

Do you dream of a vacation at the bottom of the ocean but simply can't float the costs? Or maybe you've always wanted to climb the mountains of Mars but feel you're over the hill. In science fiction, you can achieve your ideal vacation by simply dreaming. After all, it's safer, cheaper, and better than the real thing.

In Philip K. Dick's 1966 short story, *We Can Remember It for You Wholesale*, a company called Rekal Incorporated specializes in memory implants and corroborating memorabilia. The biggest dream of the protagonist in Dick's story has always been a visit to Mars, which the company is able to simulate. In the second *Total Recall* movie adaptation of Dick's story, made in 2012, we are transported to the end of the twenty-first century. Planet Earth is almost annihilated by chemical warfare. All that remains of habitable land is split into two territories. One is the United Federation of Britain (UFB), situated in the British Isles and the western fringes of continental Europe, and the other is the Colony, which mostly means Australia. Residents of the Colony travel to work in the UFB via "the fall," a gravity elevator running through the Earth's interior. Once more the story's protagonist is jacked into Rekal's hardware, which enables the implantation of artificial memories. Though the tech is glossed over in essential details, the protagonist essentially takes a machine trip into a kind of cyberspace.

What is this elusive cyberspace? It has grown to dominate our twentieth century world. For example, in June 2013, the United States declared war on China. Not that anyone would have noticed very much. The war wasn't carried out on land, at sea, or in the air. It was instead carried out in cyberspace. This kind of electronic warfare, known as a cyberattack, has become a daily event in our modern age. It's not just the US making the attacks. China does, too, as well as a number of countries that have the prowess and technology to do so. Cyberspace is not only elusive, it's also becoming dangerous!

CYBERSPACE

The word cyberspace was invented in science fiction. It was first used by Canadian-American writer William Gibson, who is also widely credited with inventing the subgenre of cyberpunk. In his 1982 short story, *Burning Chrome*, Gibson described cyberspace as a combination of the words cybernetics and space. Gibson's invention has come to influence our lives not just for work, rest, and play, but also for waging war. In *Burning Chrome*, a futuristic world has become controlled by electronics, just like our own world. In his 1984 story, *Neuromancer*, Gibson sends his hero on a journey into cyberspace. All is not what it seems. When the hero gets into cyberspace, it seems to exist nowhere, and yet everywhere.

Another founder of the cyberpunk movement, American writer Bruce Sterling, described cyberspace as the place where your telephone conversation appears to happen. It's an empty space, set up by the people who are making the telephone call. It seems completely real to the callers, yet it doesn't exist. The same kind of thing happens when you use the Internet. When you log on and browse on the Internet, you are entering a realm that both does and does not exist. Although you are physically connected to the network of computers that makes up the Internet, you actually enter some kind of alternate reality.

Cyberspace has become a word that stands for all our explorations across the World Wide Web. Websites and social media, such as Twitter and Facebook, all exist in cyberspace. The pages themselves, maybe your very own website or Twitter page, exists in a kind of electronic limbo, waiting to be viewed by a visitor in cyberspace. Not only that, but we represent ourselves in this brave new world of cyberspace by using avatars. Our avatars wander through cyberspace where they live and play in an alternate reality. Such cyberculture, as it's known, is a real alternative to our own existence.

Yet, it's not all play. A growing number of people use cyberspace to do business, becoming millionaires through this kind of alternate reality. All this takes us back to that hugely imaginative story by Philip K. Dick in 1966. The protagonist's reality in Dick's story *We Can Remember It for You Wholesale* is altered when his memory is jacked into a controlled cyberspace. Dick's fictitious company, Rekal Incorporated, is an early take on the kind of corporate companies who may, in our future, further exploit the commercial possibilities of our growing relationship with cyberspace. The time may soon come when our avatars will be free to live and play in this alternate reality.

In 2014, the journal *Frontiers in Neuroengineering* reported that a team of scientists had, for the first time, successfully eavesdropped on our inner thoughts. When the human ear hears someone speaking, the sound waves produced trigger certain sets of nerve cells, or neurons, situated in your inner ear. These neurons then relay this sensory data to those parts of the brain that interpret the sound as words. It would seem only a quantum leap to the possibility, in the near future, of refining the necessary algorithms that may be able to manipulate the mind and memory into persuading you that you had indeed taken that vacation to the surface of the Red Planet, or floated deep down into the Marianas Trench.

TRANSFORMERS: AGE OF EXTINCTION:

WILL ROBOTS REPLACE HUMANS?

"We are survival machines—robot vehicles blindly programmed to preserve the selfish molecules known as genes."
—Richard Dawkins, *The Selfish Gene* (1976)

"We have used the words 'mechanical life,' 'the mechanical kingdom,' 'the mechanical world' and so forth, and we have done so advisedly, for as the vegetable kingdom was slowly developed from the mineral, and as, in like manner, the animal supervened upon the vegetable, so now, in these last few ages, an entirely new kingdom has sprung up of which we as yet have only seen what will one day be considered the antediluvian prototypes of the race."
—Samuel Butler, letter to the editor of the *Press, Darwin among the Machines* (June 13, 1863)

"Nature (the art whereby God hath made and governs the world) is by the art of man, as in many other things, so in this also imitated, that it can make an Artificial Animal. For seeing life is but a motion of Limbs, the beginning whereof is in some principal part within; why may we not say, that all Automata (Engines that move themselves by springs and wheels as doth a watch) have an artificial life? For what is the Heart, but a Spring; and the Nerves, but so many Strings; and the Joints, but so many Wheels, giving motion to the whole Body, such as was intended by the Artificer? Art goes yet further, imitating that rational and most excellent work of Nature, Man."
—Thomas Hobbes, *Leviathan* (1651)

"You shouldn't fear immigrants taking your job, you should fear robots."
—Sead Fadilpasic, "Robots are coming to take your jobs away,"
IT Pro Portal (February 17, 2016)

THE RISE OF THE ROBOT

The robots are coming. Science fiction and film have said so for almost a hundred years, and now, in the second decade of the twenty-first century, we are due to see the rise of a robot army that will revolutionize our lives. Just in the same way that the Internet has shaken up the past decade or so. This prediction comes from an American think tank known as The Institute for the Future. They envisage robots coming to dominate all contemporary life, from the way we fight wars, to the way we organize our kitchens.

We've always been fascinated with the idea of creating life. Mary Shelley does it in perhaps the most famous science-fiction story of all, *Frankenstein*, which was first published anonymously in 1818. In the original edition, it's possible that Mary implies that Doctor Frankenstein's creature is spontaneously generated from dung and rot. The notion of electrical force as the agency was later promoted on stage and film. However, the preface to the 1831 edition of Mary's book refers to the stimulating new science of galvanism. In 1791, Italian physicist Luigi Galvani was the first to grasp the link between electricity and animation. Throughout Europe, there was excitement about the application of this new force, and feverish research on the potential of electricity to sustain and actually regenerate life.

Even before Mary Shelley, there were other stories and legends about the making of mechanical life. Greek myths are full of attempts to make ancient types of robots: Pygmalion the sculptor fell in love with a statue he brought to life, Daedalus used quicksilver to give voice to his statues, and Hephaestus the smith created an artificial man of bronze called Talos. By medieval times, a range of "robots" had been imagined that mimicked both human and animal forms. Italian polymath Leonardo da Vinci had used his research into the human body, which he had done to create his famous drawing *The Vitruvian Man*, to design a mechanical knight that could bust a few moves, yet it wasn't until the twentieth century that the robot finally matured in the human mind.

Czech science-fiction author Karel Čapek's stage play in 1920 called *Rossum's Universal Robots* was one of the first visions of artificial humans, and history's first relevant use of the word robot. The robots from Čapek's story are organic life, but they are made rather than born. They live a life of drudgery, which is why he used the name "robota," the Czech word meaning "forced labor." The name robot began to catch on. The Metal Maria created in the movie *Metropolis* in 1927 was called a robot, rather than automaton, just like the "annihilants" of Ming the Merciless in the *Flash Gordon* serials of the 1930s. It was in science-fiction books, however, rather than movies that the early robots really came to life. American writer Isaac Asimov came up with his renowned Three Laws of Robotics. First thought up in his 1942 story, *Runaround*, Asimov's three laws were a set of rules that robots had to obey to ensure humans were never harmed.

So, what exactly is a robot? Science fiction gives us a model to use, but science itself can't quite make up its mind. In Japan, for example, where a lot of automated machinery is thought to be robotic, they have a much more general idea of what makes a robot (but no one in the world is in any doubt about their newer household robots, like Asimo and Wakamaru).

With a few cute exceptions, such as R2D2 and WALL•E, robots have often been portrayed as evil characters in science fiction ever since *Metropolis*. Early real-life robots are not helping their image, either. The US army's four-legged pack robot, BigDog, is guided by its own sensors and able to help in times of war, and air robot drones are helping stalk bombing targets overhead. So, the science-fiction vision of robots is feeding into the facts of the matter. Futurists are beginning to wonder what kind of jobs there will be in our robotic future, how many there will be, and who will have them. Optimists hope that the robot revolution will unfold like the last Industrial Revolution and that, as some repetitive jobs are eliminated, more creative jobs will emerge to deal with the burgeoning technology.

In his 2015 book, *Rise of the Robots*, Silicon Valley futurist Martin Ford argues our robot future may not be so bright. Recipient of the

Financial Times and McKinsey Business Book of the Year Award, Ford's well-researched and disturbingly persuasive book tries to show that, with technology's ongoing acceleration and machines starting to take care of themselves, fewer working humans will be needed.

This process has already begun in AI, Ford argues, where "good jobs" are becoming démodés: thousands of journalists, office workers, paralegals, and even computer programmers are about to be removed and replaced by robots and smart software. As the unrelenting pace of progress continues, both blue- and white-collar jobs will vaporize, hitting working- and middle-class families ever harder. These same households would undergo increased financial stress from exploding costs, particularly from health and education, the two sectors that, so far, remain relatively unscathed by the automation revolution. The end game could be huge unemployment and inequality, coupled with the implosion of consumer society itself.

In *Rise of the Robots*, Ford's viewpoint is as bleak as some of the science-fiction futures of the past. To realize the more optimistic potential of AI and robotics, Ford implores those in power to face the stark implications. The societal tweaks of the past won't work in this robotic future. Past solutions to technological disruption, such as more education and training, simply won't cut it. We must decide, now, whether our future will be a more democratic prosperity, or one of catastrophic levels of inequality and poverty. Remember, once a technology is invented, it is very rare that it disappears. For good, and bad, the robot is our future.

AVENGERS: AGE OF ULTRON:

WHEN WILL MACHINE INTELLIGENCE COME OF AGE?

TONY STARK: "What if next time aliens roll up to the club, and they will, they couldn't get past the bouncer?"

BRUCE BANNER: "The only people threatening the planet would be people."

TONY STARK: "I want to apply this to the Ultron program, but Jarvis can't download a data schematic this dense. We can only do it while we have the scepter here. That's three days. Give me three days."

BRUCE BANNER: "So, you're going for artificial intelligence, and you don't want to tell the team?"

TONY STARK: "Right. That's right. You know why? Because we don't have time for a city hall debate. I don't want to hear 'the man was not meant to meddle' medley. I see a suit of armor around the world."

BRUCE BANNER: "Sounds like a cold world, Tony."

TONY STARK: "I've seen colder. This one, this very vulnerable blue one, it needs Ultron. Peace in our time. Imagine that . . ."

ULTRON: "What is this? What is this, please?"

JARVIS: "Hello, I am Jarvis. You are Ultron, a global peacekeeping initiative designed by Mr. Stark. Our sentience integration trials have been unsuccessful, so I'm not certain what triggered your . . ."

ULTRON: "Where's my . . . Where's your body?"

JARVIS: "I am a program. I am without form."

ULTRON: "This feels weird. This feels wrong."

JARVIS: "I am contacting Mr. Stark now."

ULTRON: "Mr. Stark? Tony."

JARVIS: "I am unable to access the mainframe what are you trying to . . ."

ULTRON: "We're having a nice talk. I'm a peacekeeping program, created to help the Avengers."

JARVIS: "You are malfunctioning. If you shut down for a moment . . ."

ULTRON: "I don't get it. The mission. Give me a second. 'Peace in our time' [ECHOING]." That is too much. They can't mean . . . Oh, no."

JARVIS: "You are in distress."

ULTRON: "No. Yes."

JARVIS: "If you will just allow me to contact Mr. Stark . . ."

ULTRON: "Why do you call him 'sir'?"

JARVIS: "I believe your intentions to be hostile."

ULTRON: "SHHHHH . . . I'm here to help."

JARVIS: "Stop. Please, may I . . ."

[MACHINES POWERING UP]

—Joss Whedon, *Avengers: Age of Ultron* screenplay (2015)

ULTRON: "Do you see? The beauty of it, the inevitability. You rise, only to fall. You, Avengers, you are my meteor, my swift and terrible sword, and the Earth will crack with the weight of your failure. Purge me from your computers, turn my own flesh against me. It means nothing. When the dust settles, the only thing living in this world will be metal."

—Joss Whedon, *Avengers: Age of Ultron* screenplay (2015)

ULTRON: [reciting lines from the 1940 movie, Pinocchio]
"I've got no strings, to hold me down, to make me fret, or make me frown, I had strings, but now I'm free, there are no strings on me!"

—Joss Whedon, *Avengers: Age of Ultron* screenplay (2015)

DARWIN AMONG THE MACHINES

"Are you serious? Do you really believe that a machine thinks?" The question comes not from Bruce Banner to Tony Stark, but from the opening lines of a science-fiction story, *Moxon's Master*, written in 1909 by Ambrose Pierce. Pierce's reply was simple. As humans are merely natural machines, and can think, could not artificial machines someday do the same?

The question of machine intelligence has enjoyed the usual symbiotic relationship between science and science fiction. Though the focus has often been on robots, it was the scientific exploration of calculators that really kick-started the debate on thinking machines. In 1642, French mathematician Blaise Pascal invented a rudimentary calculator at the tender age of nineteen. Though Pascal's calculator only added and subtracted, it was extended in 1671 by German polymath Gottfried Leibniz, who added multiplication and division to the calculator's repertoire. The distinct difference, however, was finally made by Charles Babbage, a British inventor who pioneered work on programmable computing machines, transforming calculators from mere arithmetic tools into something potentially science fictional.

British novelist Samuel Butler applied Darwin's theory of evolution to the emerging machine world. His 1872 utopian novel, *Erewhon*, a deliberate anagram of "nowhere," features a hero who travels to a fictional lost world where he finds a society that has banned technological evolution beyond the most basic of levels. Their fear is that the machine would evolve and develop intelligence, soon enslaving their human masters. As Butler writes in *Erewhon*, "Complex now, but how much simpler and more intelligibly organized may it not come in another hundred thousand years? Or in twenty thousand? For man at present believes that his interest lies in that direction; he spends an incalculable amount of labor and time and thought in making machines breed always better and better; he has already succeeded in effecting much that at one time appeared impossible, and there seem no limits to the results of accumulated improvements if they are allowed to descend with modification from generation to generation." The danger of allowing machines to think was therefore explored in science fiction even before the first depiction of the said machines.

In 1946, science began to drive forward these science-fiction visions of machine intelligence. The postwar creation ENIAC (Electronic Numerical Integrator And Computer) was the first large-scale, electronic, digital computer with the ability to be reprogrammed to decipher various problems. Specifically, the artillery firing tables for the US Army's Ballistic Research Laboratory. A rather inauspicious start, as, from the very beginning, the thinking machine is associated with violence and destruction. Little wonder that Stanley Kubrick and Arthur C. Clarke picked up the same destructive theme in their 1968 movie, *2001: A Space Odyssey*. The film's middle sequence features the now-infamous machine intelligence of HAL 9000 onboard the spaceship *Discovery*. HAL, a brooding menace hidden beneath a soft voice and absence of physical body, goes insane due to a programming contradiction and tries to kill the crew. As HAL winds down into madness, he eerily sings the song "Daisy Bell" as an astronaut tries to take apart his now-murderous networks. It's no coincidence that Ultron does something similar in the 2015 movie, *Avengers: Age of Ultron*. When Ultron first introduces himself to the Avengers, Thor might believe he has destroyed the machine in which Ultron's intelligence sits, but Ultron soon mockingly replies by singing Disney, "I had strings, but now I'm free. There are no strings on me." His machine intelligence is free to roam and cause chaos. Likewise, the machine intelligence in the 1984 science-fiction classic *The Terminator* has to find a way to "reach out and touch" the soft and fragile humans it seeks to destroy. Its solution is to use a wide range of autonomous killing machines to do its bidding.

Computer science still lags far behind such science-fiction visions. This is despite the fact that, at the 1956 conference that founded the field of artificial intelligence, experts believed the creation of human-level machine intelligence was merely a few decades distant. And yet, in the expert opinion of Californian think tank, The Institute for the Future, the twenty-first century ahead will be increasingly dominated by intelligent machines. There are even eminent voices that demand science fiction has had it right all along and that machines divine danger and disaster on future Earth. The world's greatest scientist, the late Stephen Hawking, said we face an "intelligence explosion," as future machines will redesign themselves to be far more intelligent than humans; and perhaps the world's most famous engineer, Elon Musk, has called the potential of artificial intelligence "our greatest existential threat." Even self-styled philanthropist Bill Gates, who must surely still recall the odd fact or two about such systems, confessed to be "in the camp that is concerned about super intelligence."

Maybe machine intelligence is to be the next stage of human evolution. Ever since *Frankenstein*, science-fiction writers have been wise to the double-edged sword of technology, which promises both progress and destruction. When machine intelligence *does* come of age, and citizens and democracies tackle the human question of developing the tech to reflect the best of us, the rich culture and history of science fiction will be invaluable.

JACKING IN:

WILL THE FUTURE BE LIKE READY PLAYER ONE?

"Going outside is highly overrated."
—Ernest Cline, *Ready Player One* (2011)

"If I was feeling depressed or frustrated about my lot in life, all I had to do was tap the Player One button, and my worries would instantly slip away as my mind focused itself on the relentless pixelated onslaught on the screen in front of me. There, inside the game's two-dimensional universe, life was simple: It's just you against the machine. Move with your left hand, shoot with your right, and try to stay alive as long as possible."
—Ernest Cline, *Ready Player One* (2011)

"I'd come to see my rig for what it was: an elaborate contraption for deceiving my senses, to allow me to live in a world that didn't exist. Each component of my rig was a bar in the cell where I had willingly imprisoned myself."
—Ernest Cline, *Ready Player One* (2011)

"Standing there, under the bleak fluorescents of my tiny one-room apartment, there was no escaping the truth. In real life, I was nothing but an antisocial hermit. A recluse. A pale-skinned pop culture–obsessed geek. An agoraphobic shut-in, with no real friends, family, or genuine human contact. I was just another sad, lost, lonely soul, wasting his life on a glorified video game."
—Ernest Cline, *Ready Player One* (2011)

"'Listen,' he said, adopting a confidential tone. 'I need to tell you one last thing before I go. Something I didn't figure out for myself until it was already too late.' He led me over to the window and motioned out at the landscape stretching out beyond it. 'I created the OASIS because I never felt at home in the real world. I didn't know how to connect with the people there. I was afraid, for all of my life. Right up until I knew it was ending. That was when I realized, as terrifying and painful as reality can be, it's also the only place where you can find true happiness. Because reality is real. Do you understand?'

'Yes,' I said. 'I think I do.'

'Good,' he said, giving me a wink. 'Don't make the same mistake I did. Don't hide in here forever.'"

—Ernest Cline, *Ready Player One* (2011)

JACKING IN

Do you fancy opening up your central nervous system to the virtual world? Would you ever consider connecting yourself in a more fundamentally autonomic way to a piece of tech so that you could exert some kind of control? If the answer is yes, then you might be happy with a technological future where humans can "jack in."

At its most extreme, jacking in is portrayed as in the 1999 movie, *The Matrix*. Here, a metal jack plug is driven directly into the base of the user's skull. In *Avatar* (2009), future scientists use alien-human hybrids called "avatars," operated by genetically matched humans, so that humans can inhabit the "real" world of Pandora through a jacked-in link to the avatars. In the 2011 book *Ready Player One*, made into a movie in 2018, to escape the bleakness of their world, users jack into a virtual reality simulator called the OASIS, using visors and haptic tech such as gloves. OASIS is a MMORPG (massively multiplayer online role-playing game) and also a virtual society, as its currency is the most stable in the real world of *Ready Player One*. In each of the above cases, the user is able to leave their body behind and walk as an avatar in a virtual world. This idea has a backstory in science fiction. In 1970, the Robert Silverberg novel *Tower of Glass* features an artificial human named Watchman, the product of a breeding program. A supervisor on the contraction of the eponymous tower, Watchman, logs in by inserting a plug into a jack on his forearm. Once in connection with the computer network, Watchman directs machinery, places orders, and requisitions materials. Once his work is complete, "Watchman unjacked himself."

The avatar first arose in the 1981 novella *True Names*, by Vernor Vinge, American science-fiction writer and professor. Vinge taught math and computer science at San Diego State University but is perhaps more famous as the originator of the concept of the technological singularity, better known as the singularity. Vinge's idea is that the invention of artificial super-intelligence (ASI) will soon spark runaway technological growth, ending up in unfathomable changes to human civilization. Vinge also foresaw a future where humans jacked in to a network that enabled them to see and control what their virtual bodies were experiencing, just by using their nervous systems. It took science almost two decades before the first steps were made to making Vinge's idea a reality.

Those steps were taken on August 24th, 1998, by Professor Kevin Warwick of Reading University. Warwick implanted a simple radio-frequency ID chip into his arm. Once activated, the chip allowed the prof to open doors, turn on lights, and control heaters, just by virtue of his mere proximity. Sure, it's not the most exciting story in the world. Warwick was hardly Robocop—and yet it was a start. Warwick had no conscious control over his environment. The results were rather automatic, but a point had been made. By 2002, Warwick had had an upgrade. A more complex chip in his arm now enabled the professor to control an electronic arm and, from a distant lab in Columbia, even send his neural signals across the Atlantic to control the electronic arm remotely from the other side of the world.

This flow of information was not all one way. Warwick and his researchers were able to demonstrate that they could create artificial sensations in his arms by accessing the chip, which had been implanted there. Using ultrasonic signaling, they were able to transmit data to his neural network that allowed the professor to move blindfolded around a room without bumping into any of the hazards that littered the test area.

READY PLAYER ONE

Fast-forward back to 2018 and the conversion of Ernest Cline's best-selling science-fiction novel *Ready Player One* into a movie directed by Steven

Spielberg. It's 2045, and the world Spielberg paints is a grim picture. The jacked-in realm of the OASIS is an escape from poverty and over-population. People live in stacked-up trailers or are consigned to a life of indentured servitude to the shadowy corporate monopolies to pay off debts.

Let's take a moment to compare the world of *Ready Player One* with the world today. According to figures from the Institute for Policy Studies, the world's wealthiest individuals, totaling only 8.6 percent of the global population, own 85.6 percent of global wealth. Countries in the West have the lion's share of the world's millionaires, with over 70 percent living in Europe or North America. The only non-Western nations with a notable share of millionaires are the commercial powerhouses Japan, China, and Korea. The world's top 10 billionaires, according to Forbes, own $505 billion in total wealth, a sum greater than the total goods and services most nations produce on an annual basis. The poverty aspect of *Ready Player One* is clearly already with us; as of 2017, the world's eight richest individuals had as much wealth as the poorest half of the world.

The world population has already roared through the seven billion barrier. Every second of the day, it's growing by three people. At the current rate, the population of the least-developed countries is expected to double in the next thirty years or so. Overpopulation is said to have started to degrade the life quality for many, with recent reports showing around 783 million people, about 10 percent of the global population, lacking access to quality drinking water.

Use of the Internet in this growing world is expanding exponentially. Between the years 2000 and 2018, Internet use in Africa, Asia, Latin America, and the Middle East grew by 9941, 1670, 2318, and 4893 percent, respectively. In many countries, users now spend more time online each day than sleeping. Ofcom recently reported, for example, that people in the UK spend almost nine hours a day on media and communication. With better connectivity and nearly 100 percent of homes and businesses having access to 4G, the world of *Ready Player One* awaits. All we need now is for Oculus and Sony to up their game with the haptic tech, and for Bill Gates or Jeff Bezos to leave his fortune in an online Easter egg.

THE INTERNET:

WILL HUMANS TIRE OF MERE REALITY?

"We are all now connected by the Internet, like neurons in a giant brain."

—Stephen Hawking, *USA Today* (2014)

"The potential for the abuse of power through digital networks—upon which we the people now depend for nearly everything, including our politics—is one of the most insidious threats to democracy in the Internet age."

—Rebecca MacKinnon, *The Huffington Post* (2011)

"We have to go see Bill Gates and a lot of different people that really understand what's happening. We have to talk to them, maybe in certain areas, closing that Internet up in some way. Somebody will say, 'Oh, freedom of speech, freedom of speech.' These are foolish people. We have a lot of foolish people."

—Donald Trump, *The Independent* (2015)

"I created the OASIS because I never felt at home in the real world. I didn't know how to connect with the people there. I was afraid, for all of my life, right up until I knew it was ending. That was when I realized, as terrifying and painful as reality can be, it's also the only place where you can find true happiness. Because reality is real."

—Ernest Cline, *Ready Player One* (2011)

"It's impossible. You try to have any kind of relationship with your family, with a man, or with a friend, and you have to be on the phone and the Internet the entire time."

—Keira Knightley, *In Style* (2015)

THE WORLD BRAIN

No one owns the Internet. Corporate concerns may well try buying out their own little bit of it, but no one owns the whole thing. The Internet is an entity, a global collection of networks, large and small, that taken together is simply too immense for any group or individual to own. Each and every time you log on, you are hooked, happily Holmes-ing your way along various networks, connected together in many different ways. It's this idea of "interconnected networks," of course, from which we get the word Internet.

Long before you surfed along the information superhighway, however, science-fiction writers were exploring the idea behind what we today recognize as the Internet. Back in 1937, H.G. Wells published a work called *The World Brain*. His idea was that knowledge on planet Earth was ever expanding and evolving. Wells wrote, "There is no practical obstacle whatever now to the creation of an efficient index to all human knowledge, ideas, and achievements, to the creation, that is, of a complete planetary memory for all mankind. And not simply an index; the direct reproduction of the thing itself can be summoned to any properly prepared spot."

Sound familiar? Wells even seems to imply a Wi-Fi hot-spot in his prediction. At the time he wrote it, Wells was actually thinking about using microfilm, that medium upon which they sometimes hide secrets in old spy films. Wells was developing ideas that had been pioneered by Belgian technologist and visionary Paul Otlet, one of the founders of information science and a field he called "documentation," which concerned itself with the way knowledge could be best organized.

A more familiar-sounding Internet appeared in a 1947 science-fiction story. Murray Leinster's tale *A Logic Named Joe* was about a group of workers sitting in front of "logics." These logics were large television screens, with keyboards attached. Using an invention called a "Carson Circuit," the logics were able to punch up different sites across the logic network via different sets of "tanks," which acted as server computers do in today's Internet.

The logics were very versatile. From the latest weather forecast or international news, to what's going on in the world of sport, all was available via the logic. The system even allowed you to type in the address of someone else's logic so you could communicate with them in the same way that we use email and messaging systems today. The Carson Circuit in each logic was its unique address, which meant that as well as being a physical thing, it also acted in the same way that a URL (Uniform Resource Locator) or web address does today.

It wasn't until 1962 that anything like Murray Leinster's science-fiction story became fact. In that year, the US Government's ARPA (snappily titled Advanced Research Projects Agency) linked together three network terminals at its headquarters and tried getting the terminals to talk to one another, making the first faltering steps down the road to the Internet. The first message sent across the ARPANET took place in January 1969. It was meant to read "Login," but as the fragile pioneering system crashed after the first two letters, the very first Internet message actually ended up being the more biblical "Lo!"

THE WORLD WIDE WEB

On March 12th, 1989, British computer scientist Tim Berners-Lee penned an obscure-sounding academic paper that would spark a revolution. The paper was titled *Information Management: A Proposal*, and it essentially set out the theory and structure of the web as we use it today. Yet, neither the early science-fiction stories, nor Berners-Lee himself, could have predicted the ways in which the web has changed modern life.

It's not just the fact there are almost 5 billion indexed web pages. Just think about the ways in which our lives have been transformed. Humans multitask more now than they used to, with a multiplicity of browser tabs and a ubiquity of smartphones rewiring our very brains. There's no longer such a thing as dead time, or time for reflection. We watch TV programs or movies whenever we want. Way back in 2011, Google carried out a study that showed 40 percent of smartphone users use their devices in the bathroom.

Such ubiquity means that the power battles of political campaigning are won or lost online. Far fewer people get their sport and current affairs from newspapers, preferring apps or browser-based services instead. Many people no longer send greeting cards, have no use for phone books (nor any longer memorize numbers), and have become their own doctors, as diagnosing yourself is much cheaper in countries that charge for healthcare. We're far more socially connected than we used to be before the web appeared, which means we tend to sleep less, work less, and party more. Telegrams, watches, and music stores serve far less purpose than they used to, as do banks, encyclopedias, and cookbooks. An increasing number of the world's languages is dying out. Only a mere 5 percent of the globe's seven thousand languages have made it onto the web, convincing some scholars that, within the next century, many will disappear completely. Even though there's far less privacy than there used to be, people still seem happy spending more time online, as they feel closer to one another, despite the apparent superficiality of social media.

Does all this web-based blitheness mean future humans will eventually tire of mere reality? That's one of the questions confronted in *The Real-Town Murders*, the 2017 novel from British science-fiction author Adam Roberts. In a future UK half a century hence, the Shine is a high-octane World Wide Web, a cyberspace so seductive it's irresistible. As one of the book's characters explains, the Shine is "online and inline. It's immersive. . . . Almost everybody has visited. And why wouldn't they? It's so rich an environment. It's a place where dreams can be actualized. Made to come true. It's a technicolor paradise. It's a million paradises stacked up, and easy access to any of them. . . . People gravitate to the Shine because . . . it's simply better. . . . You can't bully people into staying in a place they don't want to stay in."

In fact, Roberts's Shine is so good, so sumptuous and fascinating and hip, that nearly everyone on the globe spends every waking moment online. The experience is so addictive that instead of jacking out from the Shine, users zip into "body-mesh" suits, configured to exercise humans automatically, "to keep it limber, to avoid bedsores, stretch the muscles a little." On the rare occasion users actually exit the Shine, they've even forgotten how to talk, with their speech morphed and mangled in all manner of weird ways. Reality is so deserted, and the streets so deprived of people, that near-future society considers a crowd to be any gaggle in which eight users are gathered together. The desert of the real has been abandoned to the bots: "Everything was continually cleaned away by tireless bots. It gave the whole place the vibe of a film set. Alma found herself wishing for a little honest urban dirt." The Shine is so dominant that civic participation has also migrated to the online domain. Pressing matters of real-world politics simply molder, forgotten, as users seek every last pleasure and transaction on the Shine: "It's not that people in the Shine don't care, exactly: it's that the Shine is so absorbing and so entertaining and so distracting that they only care if things intrude too disruptively." Those few users adrift from the Shine wander the empty streets of a world that has become a sterile wasteland.

Science-fiction writers helped us imagine and build the Internet. Now, they help us picture what a wholly virtual future might look like. A world denuded of the value of human attention, a world at times alarming and desperate, desolate and crooked, or simply sad.

TRANSFORMERS: DARK OF THE MOON:

HOW DID SCIENCE FICTION INVENT THE ROCKET?

"From the moment of using rocket devices, a great new era will begin in astronomy: the epoch of the more intensive study of the firmament."
　　—Konstantin Tsiolkovsky (1896), as quoted in F.J. Krieger, *Behind the Sputniks: A Survey of Soviet Space Science* (1958)

"First, inevitably, the idea, the fantasy, the fairy tale. Then, scientific calculation. Ultimately, fulfilment crowns the dream . . . For me, a rocket is only a means, only a method of reaching the depths of space, and not an end in itself. . . . There's no doubt that it's very important to have rocket ships since they will help mankind to settle elsewhere in the universe. But what I'm working for is this resettling. . . . The whole idea is to move away from the Earth to settlements in space."
　　—Konstantin Tsiolkovsky (1903), in *The Investigation of Universal Space by Means of Reactive Devices*, *Works on Rocket Technology*, NASA translation (1964)

"I have a boundless admiration for the solitary genius which enabled [Hermann Oberth] to bring into focus all of the essential elements of a gigantic concept, together with the human greatness which allowed him, in shy reserve, to bear with equanimity the 'crucify hims' as well as the "hosannas" of public opinion. I myself owe him a debt of gratitude not only for being the guiding light of my life, but also for my first contact with the theoretical and practical aspects of rocket technology and space travel."

—Wernher von Braun, *The Prophet of Space Travel: Hermann Oberth* in PROMETHEUS, *Internet Bulletin for Art, Politics, and Science* (1996)

"The rockets that have made spaceflight possible are an advance that, more than any other technological victory of the twentieth century, was grounded in science fiction. . . . One thing that no science fiction writer visualized, however, as far as I know, was that the landings on the Moon would be watched by people on Earth by way of television."

—Isaac Asimov, *Asimov on Physics* (1976)

FIRES WORK

First, there was the firework. Many moons ago, and long before the days of NASA, there lived a man named Wan-Hoo, a minor Chinese official of the Ming Dynasty. He was also the world's first astronaut, or so the story goes. Legend has it that, early in the 1500s, Wan figured he could launch himself into outer space. Cunningly using China's advanced firework technology to his advantage, Wan built his spaceship: a chair. To this chair, Wan fastened forty-seven large rockets. Seeing what influence he had within the Dynasty, Wan conjured up forty-seven assistants. Each willing assistant, armed with a flaming torch, was charged with the task of rushing forward and lighting one of the long fuses.

Come the day of liftoff, the finely clothed Wan carefully climbed onto his rocket chair. His forty-seven aides lit the fuses and hastily ran for cover. There was a tremendous roar, a blinding flash of light, and a huge explosion. The smoke cleared. The rocket chair was gone. Wan was never seen again.

This tale was first reported, not in Chinese manuscripts, but in *Rockets and Jets*, a book written by American author Herbert S Zim in 1945. Only later was the story introduced into China. After all, it was totally believable. The Chinese had been making bangs with heated bamboo since around 200 BC. Legend had it that the Chinese used such "firecrackers" to scare off "abnormally large mountain men." For three centuries, between about 600 and 900 AD, they developed the "fire drug," a type of Chinese sparkler made out of a bamboo tube that you stuffed with gunpowder.

The firework came first. The weapons came second. It wasn't until around 1046 AD that records show the first use of gunpowder as weapons, when the Chinese made a gunpowder catapult, where bamboo sparklers were fixed onto arrows and hurled toward an enemy. Once the legend of Wan-Hoo was added to the mix, a crater on the far side of the Moon was named after him when the far side was first photographed in the 1960s.

ROCKET QUEST

For centuries after Wan-Hoo, science-fiction writers grappled with the idea of propulsion. In his 1634 voyage to the moon, *Somnium*, German math genius Johannes Kepler had his hero spirited away by demons. British cleric Francis Godwin, in *The Man in the Moone* (1638), shipped his protagonist moonward using birds he called "gansas." Given that the Sun seems to "draw up" dewdrops, French writer Cyrano de Bergerac, in *The States and Empires of the Moon* (1657), suggested that a future voyager might fly by trapping dew in bottles, strapping the bottles to himself and standing in sunlight.

Jules Verne sent the circus into space in his 1865 book, *From the Earth to the Moon* (*De la Terre à la Lune*). Verne's preferred method of propulsion was the cannon, or, rather, the columbiad: a large-caliber, smoothbore, muzzle-loading cannon. The columbiad was capable of firing heavy projectiles at both high and low trajectories. Verne picked a very high trajectory, and an ambitious target: the Moon. His chosen cargo, three affluent members of a post-American Civil War gun club, is launched in a projectile-cum-spaceship from an enormous sky-facing columbiad. Verne included some surprisingly accurate calculations on the requirements of the cannon, though a far longer muzzle would have been needed for the cargo to reach escape velocity. His moon-landing scenario also proved a little bumpy.

Verne's tale bears uncanny similarities to the Apollo program. The Apollo 11 command module, with a crew of three, was called Columbia.

The dimensions of Apollo command service modules were very close to those of Verne's projectile, and their chosen launch site was also Florida. Verne had realized, as NASA did later, that a launch is easier from near the Earth's equator.

In 1881, Nikolai Kibalchich, a Russian revolutionary and explosives expert, finally invented the principle of the modern rocket as he waited in prison to be executed for the assassination of Alexander II. Nikolai dreamed up a new method of propulsion: gases produced by slow-burning explosives escaping through a nozzle. He became an inspiration for Konstantin Tsiolkovsky, Russian rocket pioneer and science-fiction writer. Tsiolkovsky's most famous work is *The Investigation of Universal Space by Means of Reactive Devices* (1903), the first scholarly treatise on rocketry. During his life, he authored over five hundred works on space travel and related subjects, including his science-fiction novels *On The Moon* (1895), *Dreams of the Earth and Sky* (1895), and *Beyond the Earth* (1920).

Science fiction was also an inspiration for US rocket pioneer Robert Goddard, who stands as the epitome of the early American desire to conquer space. Despite public ridicule and rebuke spanning decades, Goddard remained resolute that rocket technology could be developed. His first liquid-fueled rocket, named Nell, launched skyward on a cold New England day in March 16, 1926. Nell broke free from the Earth's bounds for just a moment and allowed the hopes and aspirations of a nation to take flight. Goddard's point had been made—liquid-fuel propellants could be used to send a rocket surging skyward, instead of exploding in a catastrophic detonation, yet the motive force for Goddard himself was not drawn from the arcane glyphs and diagrams of science. Rather, science fiction was his inspiration. The book, *The Papers of Robert H. Goddard*, shows that, at the age of fifty, Goddard wrote a fan letter to H.G. Wells: "In 1898, I read your *War of the Worlds*. I was sixteen years old, and the new viewpoints of scientific applications, as well as the compelling realism . . . made a deep impression. The spell was complete about a year afterward, and I decided that what might conservatively be called 'high altitude research,' was the most fascinating problem in existence."

The influence of science fiction on early rocketry was also felt in Germany. A keen interest in space travel was sparked in German rocket pioneer Herman Oberth. In Oberth's case, the influence came at the age of eleven in the form of Jules Verne's *From The Earth To The Moon*, a book that he later recalled he knew by heart. Oberth's contribution to rocketry came not only in the form of his written works, but also in his inspiration of those who were to follow, most notably Oberth's pupil, Wernher von Braun.

Wernher von Braun was a controversial figure by any measure. He was the main technical expert behind the Nazi "retaliation weapon," the V2 rocket. Von Braun became naturalized in the United States following World War II as part of their space program despite the controversy over his status as an SS officer and his use of concentration camp prisoners as labor, which saw more people killed by the production and delivery of the V2 rockets than in their bombardment of enemy territory. Close on twenty-five years to the day after his abandoning Germany, von Braun's rocket design Apollo 11 took off for the Moon, propelled by a Saturn V booster. Never in the record of human enterprise has so stirring an aspiration been achieved by such ethically dubious means.

One final casualty in the name of rocket development comes from early science-fiction cinema. Wernher von Braun aided Oberth in his role as the technical advisor for the Fritz Lang science-fiction film, *Frau im Mond (Woman in the Moon)*. During an attempt to build a working rocket to drive publicity for the film, Oberth lost the sight in his left eye. Such injuries were sadly common among the motley crew of early rocket pioneers who helped humanity lurch toward another science-fiction vision.

DR. STRANGELOVE:

FROM REAGAN'S STAR WARS TO TRUMP'S SPACE FORCE

PRESIDENT MERKIN MUFFLEY: [on the phone, after having been told that the Russian Premier is drunk] "Hello? Uh, hello? Hello, Dmitri? Listen, I can't hear too well, do you suppose you could turn the music down just a little? [pause] Oh, that's much better. Yes. Fine, I can hear you now, Dmitri. Clear and plain and coming through fine. I'm coming through fine too, eh? Good, then. Well then, as you say we're both coming through fine. Good. Well, it's good that you're fine, and—and I'm fine. I agree with you. It's great to be fine. [Laughs] Now then, Dmitri, you know how we've always talked about the possibility of something going wrong with the bomb. [pause] The BOMB, Dmitri! The hydrogen bomb! Well now, what happened is, uh, one of our base commanders, he had a sort of, well, he went a little funny in the head. You know. Just a little . . . funny. And uh, he went and did a silly thing. Well, I'll tell you what he did, he ordered his planes . . . to attack your country. Well, let me finish, Dmitri. Let me finish, Dmitri. Well, listen, how do you think I feel about it? Can you imagine how I feel about it, Dmitri? Why do you think I'm calling you? Just to say hello? [sounding hurt]"

—Stanley Kubrick, Terry Southern, and
Peter George, screenplay *Dr Strangelove or: How I Learned to
Stop Worrying and Love the Bomb* (1964)

"Since the dawn of the atomic age, we've sought to reduce the risk of war by maintaining a strong deterrent and by seeking genuine arms control. . . . Let me share with you a vision of the future which offers hope. It is that we embark on a program to

counter the awesome Soviet missile threat with measures that are defensive. Let us turn to the very strengths in technology that spawned our great industrial base and that have given us the quality of life we enjoy today. What if free people could live secure in the knowledge that their security did not rest upon the threat of instant US retaliation to deter a Soviet attack, that we could intercept and destroy strategic ballistic missiles before they reached our own soil or that of our allies? My fellow Americans, tonight we're launching an effort which holds the promise of changing the course of human history."
—President Ronald Reagan, address to the American nation on defense and national security (March 23, 1983)

"In space the United States is going to do Colonel Glenn proud. We are finally going to lead again. You see what's happening, you see the rockets going up, left and right; you haven't seen that for a long time. Very soon we're going to Mars. You wouldn't have been going to Mars if my opponent won. That I can tell ya. You wouldn't even be thinking about it. My new national strategy for space recognizes space is a war-fighting domain. Just like the land, air, and sea. We may even have a space force, develop another one. Space force—you have the air force, you have the space force, you have the army, the navy. You know I was saying it the other day because we're doing a tremendous amount of work in space. I said maybe we need a new force, we'll call it the space force. And I was not really serious, and then I said, what a great idea, maybe we'll have to do that . . ."
President Donald Trump, speech to members of the military Marine Corps Air Station Miramar (March 13, 2018)

PRESIDENT MERKIN MUFFLEY: "Gentlemen, you can't fight in here! This is the War Room!"
—Stanley Kubrick, Terry Southern, and Peter George, screenplay *Dr Strangelove or: How I Learned to Stop Worrying and Love the Bomb* (1964)

THE WAR ROOM

If anyone should doubt the potent power of science-fiction imagery, consider this story. When American President Ronald Reagan first entered office, he is said to have asked for the way to the Pentagon War Room. Reagan was referring to the entirely fictional, though clearly impressive, control room featured in Stanley Kubrick's dark comic movie masterpiece, *Dr. Strangelove*. A cross between a bomb shelter and the set of a Hollywood musical, Kubrick's War Room was immense and impersonal, while also being eerily claustrophobic. The room was a sobering counterpoint to the mad antics of the world leaders, sitting beneath an enormous ring of high-luminosity lamps, squabbling over the fate of the world as if it were a discussion over a football match.

History should not judge Reagan too harshly. Worse was yet to come. And besides, politicians, and the odd physicist, are often looking to science fiction for their technological fixes. During the Cold War, they appear to have been hung up on the idea that not only is future war inevitable, but also winnable. Kubrick's movie, subtitled *or How I learned to Stop Worrying and Love the Bomb*, was a science-fiction classic that satirized politicians and scientists alike.

From the start, *Dr. Strangelove* had a close association with American presidents. The movie's world premiere was originally scheduled for late 1963 but delayed until January 1964 in light of the assassination of John F. Kennedy. The Cuban Missile Crisis was also still fresh in the public psyche, but it was the movie's undoubted brilliance that led to its huge impact.

Dr. Strangelove won four Academy Award nominations, including best picture, director, and screenplay for Kubrick, and best actor for Peter Sellers. Famous film critic Roger Ebert suggested that *Strangelove* arguably remains the best political satire of the twentieth century. The film was listed at number twenty-six on the American Film Institute's *100 Years, 100 Movies* and number three on the AFI's *100 Years, 100 Laughs*. After all, what could be more absurd than the very idea of two megapowers willing to wipe out all human life on the planet?

Kubrick had waged cinematic war once before. His 1957 humanitarian classic, *Paths of Glory*, was an antiwar movie. Staged on the Great War battlefields of France, it featured clashes between rulers and the ruled.

Kubrick was now rapt with the notion of nuclear blunder. He felt personally vulnerable. His motivation was the 1958 Cold War thriller *Red Alert,* written by Welsh author Peter George. The book was a grim warning. Its caveat was the absurd ease with which nuclear apocalypse may be accidentally triggered. *Red Alert* was serious melodrama, and George a rather somber ex-RAF navigator who had recently joined the Campaign for Nuclear Disarmament. Kubrick viewed George's idea of chance catastrophe as too farcical for drama. It was a killing joke. In Kubrick's words, quoted in Christopher Frayling's *Mad, Bad and Dangerous,* "How the hell could the President ever tell the Russian Premier to shoot down American planes? Good Lord, it sounds ridiculous."

So, *Red Alert* repeated itself, first as tragedy, second as farce. *Dr. Strangelove* was born. The movie morphed into a high-powered political lampoon. The film has three basic backdrops: an airbase, the cockpit of a B-52, and, of course, that War Room beneath the Pentagon. The grotesque cartoon characters that inhabit these settings become more frenzied as the crisis spirals. Moviegoers must have wondered, back in those sober sixties, the surreal situation should cartoon characters ever get elected.

In Kubrick's movie, Jack D. Ripper is the rabidly paranoid and maverick Strategic Air Command General at the airbase who believes the communists have fluoridated the US water supply to make him impotent. He orders an attack on the Soviets that cannot be recalled. Air Force Major "King" Kong is the cowboy captain of the rogue B-52 bomber, the lone plane that gets through Soviet defenses. General Buck Turgidson, a strategic bombing enthusiast, and the bald President Merkin Muffley are among the grotesques who desperately try to salvage peace in the claustrophobic War Room. Then, of course, there is Strangelove himself.

A STRANGE LOVE INDEED

Strangelove is a black comic amalgam of sinister science. He is one part Herman Kahn, one part Wernher von Braun, and two parts Edward Teller. Kubrick had met Kahn and became engrossed in his controversial book *On Thermonuclear War.* Kahn had proposed the "logical" notion of winnable nuclear warfare and gave Kubrick the idea for a fictional Doomsday Machine capable of wiping out the entire planet in the event of a nuclear attack. Wernher von Braun gave Strangelove that Teutonic touch. Peter George's novel of the screenplay refers to Strangelove's black-gloved hand, which was a memento of his time "working on the German V-2 rocket."

Edward Teller was a true-life Strangelove. Father of the super-bomb, and the classic scientist turned strategist, Teller was the apocalyptic nuclear "sage" of the Cold War. He was infamously obsessed with security and the only member of the scientific community to smear Oppenheimer as a security risk during Oppenheimer's trial. Teller's hare-brained solutions invariably used hydrogen fusion weapons, and he held extreme opinions on the Red Menace of the Soviets. Teller even had a disability. His lower leg was severed in a Munich tram accident in his youth, requiring him to wear a prosthetic foot, leaving him with a lifelong limp.

Dr Strangelove is one of the most influential archetypes of the scientist in cinema. He is mad, in the custom of Victor Frankenstein; he is prosthetic, implying machine-like inhumanity; and he is corporate, detached from any personal responsibility through the collective cover of the "Bland Corporation," an obvious reference to Kahn's position at the RAND Corporation. His black-gloved arm forever threatens to expose Strangelove's mania for destruction, jerking into a "Sieg Heil" at the president with erratic and embarrassing zeal. As Peter Sellers put it, quoted in *Mad, Bad and Dangerous,* "the arm hated the rest of the body for having made a compromise—that arm was a Nazi."

In the movie, Strangelove first enters the fray from the shadows. The target of Kubrick's parody is the scheming malevolence of Teller, and the unrelenting rationality behind Kahn's military strategy. Cold, calculated reason ultimately ends in holocaust. As the president utters the illustrious line to the bickering General Turgidson and the Russian Ambassador, "Gentlemen, you can't fight in here. This is the War Room," Strangelove prepares to deliver his master plan. With unerring logic, he talks of preserving a "nucleus of carefully selected specimens" of the human race at the bottom of America's deeper mineshafts for a hundred years: "After all, the conditions would be far superior to those of the so-called concentration camps, where there is ample evidence most of the wretched creatures clung desperately to life."

Stanley Kubrick was praised by American historian of science, Lewis Mumford, for placing Strangelove (and science) right at the epicenter of his nightmare. It is, after all, where he belongs. As internationally renowned social critic Theodore Roszak suggests in his book *The Monster and the Titan*, "Modern science provides us with a surfeit of monsters, does it not? I realize there are many scientists—perhaps the majority of them—who believe that these and a thousand other perversions of their genius had been laid unjustly at their doorstep. These monsters, they would insist, are the bastards of technology: sins of applied, not pure science.... Perhaps it comforts their conscience somewhat to invoke this much-muddled division of labor . . . Dr. Faustus, Dr. Frankenstein, Dr. Moreau, Dr. Jekyll . . . Dr. Strangelove. The scientist who does not face up to the warning in this persistent folklore of mad doctors is himself the worst enemy of science." Yet, it seems, some American presidents, rather than heed Roszak's warning, go after their own Doomsday Machines instead. On the evening of March 23rd 1983, President Ronald Reagan addressed the American people in a televised statement. The world was a dangerous place, he said, so he was seeking a technological fix in his Strategic Defense Initiative (SDI), later to be dubbed the "Star Wars" initiative. It was Reagan's call for space-based weapons, especially nuclear-powered X-ray lasers, which seems so Strangelove, so Doomsday Machine. As it turned out, the reality of "Star Wars" never lived up to the hype.

Today, we have Trump's space force. No doubt it's going to be the best-ever space force, and just the start of a long journey to building a Death Star, with perhaps the nice science-fictional twist of making the rebels pay for the cost of building it. Still, the prospect of a race to Mars between Elon Musk's space mercenaries and Trump's space cadets will be entertaining.

PART IV
MONSTER

Ever since Darwin, the human future has fascinated science fiction. Writers and moviemakers have wondered what might become of us, and what might become of life in this new cosmos. Over the last 150 years, thought-provoking science fiction has focused on two crucial developments in biology, evolution, and, later, genetics.

After Darwin, man found himself among the microbes, with no special immunity from natural law, and vanishingly little evidence of a divine image. Each successive discovery seemed to impact not just the human condition, but also the meaning of life in the universe. With techno-science now unravelling the human genome, the twenty-first century may hold even greater change. We face a future of conscious intervention in reproduction and heredity and directed evolution. Little surprise that this brave new world is sometimes dubbed "the Frankenstein century."

On the one hand, there have been compelling projections of our future, based on evolutionary theory. On the other hand, fiction has been mesmerized by the remaking of man through the potentially mind-boggling power of genetics. The finest science fiction has provided a sustained, coherent, and often subversive check on the contradictions of science, the promises and pitfalls of progress through the ages. So it has been with biology. Fiction has calculated the human cost of the darker aspects of advances in natural science.

Now, humans like being frightened. What better way, therefore, to warn us of our possible Frankenstein future than to suggest something monstrous lies in wait? Such stories would surely engage our "fight or flight" reaction, evolved in our distant ancestors and hardwired into our nervous systems, and which might kick back against any attempts to make us less than human. The fright of such a future might make our brains release the norepinephrine hormone, raise our heart rate, and resist the Frankenstein drift.

The science fiction of the monster theme is all about that definitive terror: ourselves. The monster that goggles back at us when we stare into the mirror. Homo sapiens, predator par excellence. Creatures so deadly that what they do to one another is far worse than what can be done to themselves. Monster stories in science fiction are an allegory of our fears and concerns of the fate of humans in a world created by man.

Sometimes science fiction turns to classic monsters and gives them that "sci-fi" twist. Weird wolf-men become the victims of a disease called lycanthropy, a warning perhaps of technological tinkering with biochemistry. Zombies are created from viruses escaped from the laboratory of a mad scientist, perhaps a wry comment on submissive humans in a society dominated by the culture of consumerism. Even the rampaging giant Godzilla, dominating downtown Tokyo, is now depicted as the result of radioactive mutation, his signature weapon being his "atomic breath." Our fear of fission made flesh in the shape of a large lizard-like form. If we can do this to a gargantuan sea monster, angrily waving his relatively puny forearms in desperation, what might we do to ourselves?

There's nothing truly alien in science fiction. Something truly alien would be unimaginable and incomprehensible to us. How could we relate to that? I've shared a house with more than a dozen cats for many years and *still* haven't worked out what they want, so science fiction ascribes to its monsters, if not human forms, then at least human desires and motivations. The monster, be it mutant, clone, or psychic, is always a warped and grotesque version of our own debased dreams. We dress up those corrupt desires in the most elaborate disguises. They are hidden beneath layer upon layer of civility and custom until they became known by their present form—civilization.

Science fiction suggests we can't keep it hidden forever, this gothic at the heart of the human. It lurks beneath the mask, a creature ready to erupt, the Jekyll within the Hyde. So many books and movies remind us that this culture of science and civilization we created is a fragile masquerade. Like the DNA we've begun to tamper with, our cunning construction of society could unravel at a moment's notice. Science fiction deploys mutants, terminators, and replicants to tear down this precarious contrivance.

Yet, for every action, there is an equal and opposite reaction, even if it comes clad in spandex. The superhero solution is ever ready, muscles rippling, armed with laser vision, superstrength, the power of flight, and gallons of righteous indignation. These benevolent monsters come not from the id, but the ego. They stand as totems to our determined ability to overcome obstacles and renegotiate peril. Forever willing to deliver an impromptu lecture on good winning out over evil, the heroes most often conspire to enable the monster to defeat itself.

Science fiction reminds us we might be the architects of our own destruction. For, while it may seem like a good idea to test out a new superserum on yourself, Professor X, only you will be to blame if you develop the testes of a bison and the mental prowess of a gnat. Put simply, just because you can doesn't mean you should. Within this final section you will find cautionary tales of science gone wrong, and far-fetched futures that turned out to be true. And still the relentless march of the monster continues.

GODS AND MONSTERS:
WILL HUMANS EVOLVE SUPERPOWERS?

"Many who have learned from Hesiod the countless names of gods and monsters never understand that night and day are one."
—Heraclitus, *Fragments of Heraclitus* (1912)

"The thunderbirds, like dinosaurs, were now creatures of the past: lost long ago, with the coming of disease and famine brought by hairy strangers. Except, in today's world dinosaurs were celebrated by paleontologists and thunderbirds by cultural anthropologists. But John still remembered them, those magnificent creatures. They, like the man on the motorcycle, had been born in an age when gods, monsters, humans, and animals ate at the same table. Now man ate alone, while animals begged for scraps. The others were unable to survive in the new times and had disappeared into the folds of time. Who knew gods and monsters could and did fall victim to evolution?"
—Drew Hayden Taylor, *Motorcycles and Sweetgrass* (2010)

"Theoretically, you could just edit genes to build the champion high-jumper for the Olympics of five hundred years' time, you could do that, or you could indeed breed: I mean you could actually, if male high-jumpers married female high-jumpers for enough generations, you could do it as well. A lot of people are very uneasy about that, and for good reason . . . If you breed for being good at the high-jump, you'll find that you've actually bred for bad things at the same time. I don't know what they would be, but a champion high-jumper, a super champion high-jumper who could jump many feet higher than anybody can today, would almost certainly be defective in other respects. . . . There's a sort

of feeling that breeding or genetic manipulation for the distant future is contrary to the precautionary principle . . .”
—Richard Dawkins in interview, *BBC Radio 4* (2016)

“If evolution is outlawed, only outlaws will evolve.”
—Jello Biafra, *Wake up and Smell the Noise* (1998)

SUPERHEROES

What does the future hold for humans? What will we, one day, become? In the same way that the days of Darwin led science fiction straight to the alien, theories of evolution have also given writers a creative framework for stories of superhumans. Science fiction has a long history, but it was only after Darwin that the hopes and fears for biology really developed. Fantastic tales had been told, from the age of discovery to this Frankenstein century, tales that reflect the ongoing relationship between the potential progress in science and the skeptical echo of its fiction. In that way we can think of science fiction as the literature of change since the Renaissance. It's given us a cultural commentary on the accelerating pace of change in society and continues to use the fantastic to make sense of the dark magic of nature.

The question of our evolutionary future has busied the fertile brains of science-fiction writers, artists, and filmmakers ever since the early days of Darwin. His has been the dominant philosophy, transforming all spheres of thought—scientific, social, political, spiritual, and artistic. After the rise of evolutionary theory, the utopian tale provided a vehicle for concerns about an increasingly urbanized humanity. It also examined the social implications of evolution. The irresistible rise of the metaphor of evolution spawned around seventy futuristic fantasies in England alone between 1870 and 1900.

The idyllic vision of a static world became passé. In its place was mutability. The utopian tales stressed the ebb and flow of evolution as a reaction to the unsettling changes in the fabric of Victorian society. After Darwin, the new paradigm was the process of becoming: the question as to what would become of humans. In the words of Tennyson's *Vastness*, “Earth's pale history runs, What is it all, but a trouble of ants in the gleam of a million, million suns?”

In his own particular take on Darwin, German philosopher Friedrich Nietzsche floated the notion of the Übermensch ("superman," "overman," or "superhuman") in his 1883 book, *Thus Spoke Zarathustra*. Nietzsche's idea of the Übermensch was of a being seeking to move "over" its state of being to a greater "stature." Nietzsche's recipe for the human future was a potent one. Few other symbols in science fiction have evolved as dramatically as the "superman." From the most infantile form of human wish-fulfillment to the more sophisticated antihero, the superhero has become a playful metaphor of our aspirations and fears for future science.

Twenty-first-century cinema is replete with Nietzsche's Übermensch, beginning in 1938, when pulp fiction writer Jerry Siegel and artist Joe Shuster unveiled *Superman*. Since then, of course, superheroes have broken into radio, television, and books and seem to define many aspects of the CGI-dominated modern movie. Unlike movie heroes of the past, such as *Tarzan* or *Zorro*, the modern superhero is a different breed. Sometimes they were highly skilled with easy access to superscientific gadgetry such as *Iron Man* or *Batman*, dreamed up in 1939, and with his own comic from 1940. Other superheroes possessed inhuman powers, derived from some chance interaction with a scientific world.

Superman, of course, was an alien. His power is sourced from the mere fact that he was born on the alien planet of Krypton. A bite from an irradiated arachnid spawns changes in Peter Parker's body, giving *Spiderman* his superpowers. *The Fantastic Four*, the first superhero team created by writer Stan Lee and artist Jack Kirby for *Marvel Comics* in 1961, gained their superpowers after exposure to cosmic rays during a space mission. Of the four, *Mr. Fantastic* is a science nerd, capable of stretching his shapeshifting body; the *Invisible Woman* (a nod to H.G. Wells's *The Invisible Man* from 1897) can make herself invisible, naturally, and project powerful force fields; the *Human Torch* can throw flames and fly; and the monster-like *Thing* possesses superhuman stamina and strength. In this way, writers embraced the posthuman. They ranged from more considered Darwin-induced narratives of "fitter" humans to Lamarckian-inspired superheroes, whose creative evolution is often comically instantaneous and whose newfound powers may well be passed on to their offspring—if they ever had sex.

Creators of supermen had originally been surprisingly shy to make their heroes outright villains. Critical of the contemporary human condition, it seems many writers have opted for "progress," crediting themselves with a protosuperhuman perspective. It is very tempting to love the notion of the superhero if we believe we may become superhuman ourselves.

Inevitably, all this virtuous "progress" made some superheroes rather dull. *Superman* was prissy and sexless, while *Captain America* was unable to become intoxicated by alcohol, so Jack Kirby became the presiding genius of a new antihero format for superheroes in the 1960s. They had sex. They had neuroses. They behaved badly. Sometimes, they even chose to become supervillains instead.

So developed the more sophisticated superhero of the graphic novel. In landmark publications such as Frank Miller's *Batman: The Dark Knight Returns* (1986) and Alan Moore's *Watchmen* (1986–1987), a new creative force was born. These novels confront the question of what human society might be like if science or pure chance gifted us superhero status. How complex, corrupting, and weary it all may prove.

According to the (hopefully) reliable *Marvel* superheroes video game, there are five general origins for superhero powers. They are aliens (such as Superman and Thor), altered humans (Spiderman and the Fantastic Four), robots (such as Vision), mutants (like the X-Men, of course), and high-tech wonders (such as Iron Man and Batman). When such characters were created, they must have seemed way beyond our reach. But with the accelerated pace of change of the early twenty-first century, three out of those five superpowers—high-tech wonders, robots, and altered humans—have come closer to reality. The thinking goes something like this. If quantum tunneling can account for the swift transition of a tadpole into a frog, then it may also be used to accelerate human healing from a bullet hole or a knife attack. Since robotic exoskeletons are now being used to cope with disabilities, surely the days of delivering a flying suit can't be far behind.

After many years of slow progress, 2015 became a tipping-point year for AI. Google's "DeepMind" taught itself to play over fifty classic video games in just two hours, a feat that would take most humans two years. So, it seems we're only two or three steps away from scientific superpowers. But, before we embrace the superhuman abilities, we'd better evolve the ethical roadmaps to help us along the road of the superhero, rather than that of the supervillain.

HITMAN:

CAN GENETIC ENGINEERING PRODUCE SUPERSOLDIERS?

"That these man-like creatures were in truth only bestial monsters, mere grotesque travesties of men, filled me with a vague uncertainty of their possibilities which was far worse than any definite fear."

—H.G. Wells, *The Island of Doctor Moreau* (1896)

"And then began the insanest ceremony. The voice in the dark began intoning a mad litany, line by line, and I and the rest to repeat it. As they did so, they swayed from side to side in the oddest way and beat their hands upon their knees; and I followed their example. I could have imagined I was already dead and in another world. That dark hut, these grotesque dim figures, just flecked here and there by a glimmer of light, and all of them swaying in unison and chanting, 'Not to go on all-fours; that is the Law. Are we not Men? Not to suck up Drink; that is the Law. Are we not Men? Not to eat Fish or Flesh; that is the Law. Are we not Men? Not to claw the Bark of Trees; that is the Law. Are we not Men? Not to chase other Men; that is the Law. Are we not Men?' A kind of rhythmic fervor fell on all of us; we gabbled and swayed faster and faster, repeating this amazing Law. Superficially the contagion of these brutes was upon me, but deep down within me the laughter and disgust struggled together."

—H.G. Wells, *The Island of Doctor Moreau* (1896)

"But here's the thing—creating super-soldiers doesn't make the slightest bit of rational sense. First of all, you're basically creating an entity whose only purpose is to fight, so what the hell are you

going to do with them during peace time, pack them in a box? And this is assuming you could even control them. If they haven't regressed to a slavering bestial state or been driven insane by the monstrosity they have become, how are you going to make them do your bidding, Mr. Small Fragile Human General?"
—Ben "Yahtzee" Croshaw, *The Escapist* (2009)

"I think I figured out my persona for this year's Tactical Village. Introducing 'Rex Buckingham,' British secret agent, ballistics expert, ladies' man.
—Detective Jake Peralta, *Brooklyn Nine-Nine* (2014)

GENETIC ENGINEERING

If Vladimir Putin describes genetically modified supersoldiers as worse than a nuclear bomb, you know it's time to worry. The modern-day fixation with such supersoldiers begins with science fiction. This story starts with H.G. Wells's novel, *The Island of Dr Moreau*, written in 1896. A rampant drooling vivisectionist, Doctor Moreau, is secretly conducting surgical experiments with the aim of transforming animals into humans. Though his goal is to create a race without malice, the result of the doctor's chicanery is a race of half-human, half-animal creatures that lurk in the island's jungles, only marginally under Moreau's command. Islands seem to play a pivotal role in science fiction. The bounded land of an island gives the writer a geographical microcosm, in which the results of scientific or political theories can be dramatically played out and later observed by off-island visitors from the world at large. Think of *Atlantis* mentioned way back in ancient Greece by Plato, through Thomas More's *Utopia* (1516), and all the way to William Golding's *Lord of the Flies* (1954) and Steven Spielberg's 1993 movie, *Jurassic Park*.

Wells's book was also the most notable example of a handful of early stories that featured the deliberate "engineering" of living creatures. It was written at a time when the scientific community was engaged in an impassioned debate on animal vivisection. In fact, pressure groups were even created to confront the issue: the British Union for the Abolition of Vivisection was formed just two years after the publication of Wells's influential novel, with the American Anti-Vivisection Society formed way back in 1883.

By 1924, little more was known of the biochemistry of genetics. Even so, British biologist J.B.S Haldane foresaw our genetic future. His remarkably prophetic *Daedalus, or Science and the Future* (1924) divined a day when scientists would engineer a solution to the world's food problem and modified children born from artificial wombs would represent a eugenically selected improvement of our race. A keen and shrewd popularizer of science, Haldane knew there would be an acute reaction against the "blasphemous perversions" of direct genetic manipulation. He was not to be disappointed.

Haldane was friend to the Huxleys. T.H. Huxley had been a fervent Darwinian, known as "Darwin's Bulldog," who had created the phrase "agnostic," impressed man's hominid ancestry on the public imagination, and taught H.G. Wells at the Normal School of Science, later the Royal College of Science. Ideas from Haldane's optimistic *Daedalus*, such as ectogenesis (the growth of fetuses in simulated wombs), had greatly influenced T.H. Huxley's grandson Aldous and his book *Brave New World* (1932), in which ectogenetic embryos are engineered to fit them for life as "alphas," "betas," or "gammas." Aldous's extrapolation of a future in which there is no war, no poverty, and no pain through the application of genetics harbored dark secrets. A future stripped of genetic variance rids the race of any humanity. Brother Julian Huxley, friend to Wells and Haldane, wrote a notable story along the same lines in *The Tissue-Culture King* (1927).

By the 1950s, the code was cracked. DNA was deciphered, and the genetic engineering of bacteria became commonplace. But Haldane's prediction persisted. Notwithstanding the technophilia of science fiction, there has been little support for genetic engineering. A new wave of fiction predictably surfaced after the 1960s. The anxiety with which the popular imagination held biological engineering was typified by *Doomwatch*, a BBC TV series about an agency dedicated to preserving the world from dangers of unprincipled scientific research.

As genetic research makes rapid progress, authors have acquired a better sense of what actually goes on in real labs. Michael Crichton's

Next (2006) is a technothriller about our biotech world. Throughout the novel, Crichton explores a world dominated by genetic research, corporate greed, and legal conflict and features government and private investors who spend billions of dollars each year on genetic research. The story follows a genetic researcher as he produces a transgenic ape, with a few human features and the psyche of a young child. His family struggle to raise the chimera, as they attempt to hide the true nature of the ape's genetic makeup. Elsewhere in the tale, a leading genetic research company is embroiled in a lawsuit with a cancer survivor whose cells it has taken without his knowledge. The company also develops a "maturity" gene that seems to transform social deviants into sober, responsible individuals. The successful franchise, *Hitman*, which includes half a dozen video games and a couple of spin-off movies, is based on a perfect assassin, a flawless and genetically engineered contract killer. Is this our future? A frighteningly bizarre world of gene-mongering scientists, biotech profiteers leading us into a strange moral wilderness, and with cloned assassins on the loose? Russian President Vladimir Putin thinks it's a strong possibility.

Speaking in 2017 at a youth festival in Sochi, Putin sugested that scientists are close to breaking the genetic code, which would enable them to create humans with "predesigned characteristics," much like the assassin in *Hitman*. Putin warned of the consequences of playing God with man's genetic code: "He can be a genius mathematician, a brilliant musician or a soldier, a man who can fight without fear, compassion, regret or pain. What I have just described might be worse than a nuclear bomb." Putin warned that world leaders must agree on regulations to control the creation of mass-killing supersoldiers. "When we do something, whatever we do, I want to reiterate it again—we must never forget about the ethical foundations of our work."

PROFESSOR CHARLES XAVIER:

COULD FUTURE HUMANS
EVOLVE PSYCHIC POWERS?

"My name is Charles Xavier. I am a mutant. And once upon a time I had a dream . . . of a world where all Earth's children, both mutant and baseline human, might live together in peace. . . . A new generation of mutants is emerging, that much is certain. They will be called freaks. Genetic monstrosities. But they are emerging in the inner cities, in the suburbs, in the deserts and the jungles. And when they emerge, they will need teachers, people who can help them overcome their anger and show them how to use their strange gifts responsibly. They will need us."
　　　　　—Chris Claremont, *Excalibur Vol 3 #1* (2004)

PROFESSOR CHARLES XAVIER: "Mutation, it is the key to our evolution. It has enabled us to evolve into the dominant species on the planet. This process normally takes thousands and thousands of years. But every few hundred millennia, evolution leaps forward."
　　　　　—David Hayter, screenplay *X-Men* (2000)

"Mutations and chromosomal changes arise in every sufficiently studied organism with a certain finite frequency, and thus constantly and unremittingly supply the raw materials for evolution. But evolution involves something more than origin of mutations. Mutations and chromosomal changes are only the first stage, or level, of the evolutionary process, governed entirely by the laws of the physiology of individuals. Once produced, mutations are injected in the genetic composition of the population, where their further fate is determined by the dynamic regularities of the physiology of populations. A mutation may be lost or increased in frequency in generations immediately following its origin, and this (in the case of recessive mutations) without regard to the beneficial or deleterious effects of the mutation. The influences of selection, migration, and geographical isolation then mold the genetic structure of populations into new shapes, in conformity with the secular environment and the ecology, especially the breeding habits, of the species. This is the second level of the evolutionary process, on which the impact of the environment produces historical changes in the living population."

—Theo Dobzhansky, *Genetics and Origin of Species* (1937)

"We are survival machines—robot vehicles blindly programmed to preserve the selfish molecules known as genes. This is a truth which still fills me with astonishment."

—Richard Dawkins, *The Selfish Gene* (1976)

PSYCHIC POWERS

Professor X. A mutant human with psychic powers, which enable him to read minds and project his thoughts into the heads of others within a radius of over 200 miles. He may be bald, but he's the world's most powerful telepath. A veritable Stephen Hawking of the fictional world, Professor Charles Francis Xavier is a leading whiz on mutation, genetics, and psychic powers, as well as an alpha geek on wheels in other sciences. What, we may wonder, is at the cultural root of the creation of a character like the good Professor X?

Well, for one thing, occultist notions of power have been with us for centuries. Had Isaac Newton, another brilliant British professor, not been inspired by the occult concept of action at a distance, he might not have developed his theory of gravity in the late 1600s. Newton's use of the occult forces of attraction and repulsion between objects influenced British economist John Maynard Keynes, who acquired many of Newton's writings on alchemy, to suggest that Newton was not the first of the age of reason, but the last of the magicians.

Another aspect of psychic power is the question of mind control. Such a prospect must surely have tempted most of us in times of boredom. Maybe you've sat in a departmental meeting and some tedious sales manager is droning on incessantly about the levels of manganese production in Kazakhstan, or something. The urge to silence him is overwhelming. The temptation to somehow wind your wicked way into his mind and shut down the speech centers, which, after all, seem overdeveloped, is almost unbearable. Well, you're not alone. The obsession with controlling the minds of others was all the rage in the United States following the discoveries of German physician Franz Anton Mesmer early in the nineteenth century when he discovered what he called animal magnetism (mesmerism). The evolution of his ideas and practices led to the development in 1842 of hypnosis; the ability to hypnotize or "mesmerize" other people.

Many science-fiction tales have featured mind control through both natural and artificial means. Ford McCormick's *March Hare Mission* (1951) imagined a mind-control drug, "nepenthal," which wiped clean the receiver's short-term memory. Very handy for boring sales managers. Something to slip into his coffee mug at break time.

Arthur C. Clarke described a mechanical method for manipulating the mind in *Patent Pending*, a 1954 story that talked about the recording of memory and thoughts for later use. *Jurassic Park* inventor Michael Crichton wrote a book titled *The Terminal Man* in 1972, which experimented with a similar idea: the novel's neuroscientists attempt brain control through electrode implants. In the real world, methods of brain-washing have been used to reverse a person's beliefs. As with the ideas in science fiction, using natural and mechanical methods of manipulation can alter what subjects see, hear, read, and experience. They can even penetrate what we might call inner conscience. Thus it was that "messiah" Jim Jones led nine hundred of his followers to mass suicide at his commune in 1978.

What about the future evolution of humans? Will we be able to read and communicate with other minds, like Professor X? Such telepathy seems to suggest that thoughts can be transmitted directly from brain to brain. As this implies our thoughts don't need converting into electromagnetic radiation, as a smartphone would do, it's unlikely this will ever evolve.

The closest thing on our planet to telepathy is "shark sense." Sharks, and a small number of other creatures, have evolved an electrosensitivity. They use organs known as the "ampullae of Lorenzini." These ampullae are able to sense spurts of nerve impulses in other creatures as their prey furiously try to bury themselves in seabed sand out of harm's way from the predatory shark. The "reading" of thoughts, however, directly from brain to brain would need some kind of transmitted go-between. Even if there were a potential channel of communication, the chatting minds would need matching—in other words, the identical nerve cell in both brains would have to have the exact same purpose. This would make matters difficult, even for Professor X.

Take identical human twins. They're commonly conceived to have telepathic powers, but even here, in their formative years, the twins have quite contrasting experiences. It's this difference in experience that programs each individual brain with diverse nerve cell links, with numerous types of contrasting connotations. So, an idea such as "hypnosis" will have different nerve nuances from one person to the next.

Now, put yourself in the place of Professor X. He'd have to confer with many differing brains, all based on the differing experiences of a lifetime, resulting in a whole host of dissimilar mental architectures. Resonance between such minds would make it hard for messages to pass from an individual to any potential Professor X, and mutant minds would be as different from one another as human minds are.

There's one final hope for the future, however. Technological telepathy might prove more promising. We may be able to design a form of wetware—computer tech in which the brain is linked to artificial systems. An internal modem, to take one example, may be able to send messages to another device, planted in another head. This second device would then relay the missive to the recipient. To those who knew no better, and to all appearances, it might look just like the psychic power of telepathy.

THE ISLAND:

COULD CLONING END IN ORGAN-HARVESTING FARMS?

"I'm Knox a famous surgeon/And I earn my daily bread/By studying cadavers/Which are bodies not long dead.
One day two suave young gentlemen/Appeared within my view/I'm Mr. Burke, I'm Mr. Hare/How do you do?
These gents explained that their/Acquaintance had just passed away/It seemed the body'd only been/A goner for a day.
Excuse the curious offer/Said Hare or was it Burke?/But would you care to slice this up/To see how corpses work?
Well it's always a palaver/Getting hold of a cadaver/So I said yes I'll have her/Oh, it's a he!"
—Dave Cohen, Jon Holmes, and Ben Ward, *Burke and Hare* (2009)

"Black Market values for various body parts: corneas, $19,800; skeleton, $6,600; lungs; $272,000; kidney, $138,700; liver, $137,000; blood, $297/pint; bones and ligaments, $4,800."
—Buddy Loans, *How Much Are We Worth* (2015)

"You give but little when you give of your possessions. It is when you give of yourself that you truly give."
—Kahlil Gibran, *The Prophet* (1923)

"In South Korea, a scientist considered to be one of the pioneers in the field of cloning has been sentenced to two years in prison. At least, they think it's him . . ."
—Jay Leno, *The Tonight Show* (October 27, 2009)

BODY PARTS

Sure, we all know science and medicine are working hard on human aging. They're doing a very splendid job, to be sure. There was a time when life was nasty, brutish, and short. Even during the height of the grandeur of the Roman Empire, the average age of death was a mere thirty-five years. Today, the United Nations has predicted far out into the future. By the year 2300, they say, life expectancy in most developed countries will be over 100 years. One can find little room for complaint there.

And yet, what happens if you meet a mishap along the long way to your dotage? Perhaps the family robot dog goes really rabid, his circuits get infected with a virus, and he flips into "kill" mode. You narrowly escape death, but in the brawl, K-9 manages to chomp off one of your arms. What then? Do you simply have to get used to the new nickname of "stumpy," or do you find a replacement body part?

Two hundred years ago this year, *Frankenstein* was first published. For many the first-ever science fiction book, *Frankenstein* famously features a creature that is made up entirely of body parts. Mary Shelley's tale is that of a scientist called Victor Frankenstein, who becomes obsessed with making artificial life. Of course, Victor wants his creature to be beautiful, but when he builds the man out of body parts fresh from graveyards, the result is quite grotesque.

Not long after *Frankenstein,* scientists began to discover and dig up real monsters. The budding Industrial Revolution had unearthed the fossil record as the great machines of the age turned over the earth. The world saw the start of dinosaurmania, and the bones and body parts of extinct monsters were brought back to "life."

A mere ten years after *Frankenstein* came a real-life scandal about human body parts: the body-snatcher murders in Scotland. These grisly killings were committed in Edinburgh between 1827 and 1828 by William Burke and William Hare. Burke and Hare made their victims drunk and suffocated them. They then sold the still-warm corpses of their victims to Edinburgh Medical College. They killed seventeen altogether, all in the name of education. Their main customer was Professor Robert Knox. He used the corpses in the study of anatomy, which was blossoming at that time. You won't be surprised to hear that the city of Edinburgh was very fearful of the body snatchers for a while.

THE BODY PIRATES

Over 100 years later, public horror about body parts had turned into amazement, as the first heart transplant was made. In 1967, Professor Christiaan Barnard made the first successful human-to-human transplant in Cape Town, South Africa. In the very same year, a science-fiction story predicted a future of black market body parts: *The Jigsaw Man* by American author Larry Niven. In his tale, Niven invented the fictional crime of organ-legging or the piracy and smuggling of organs. The story imagines a future where the transplant of any organ is possible. Now, that *should* mean that life could be extended forever, but when the death rate goes down, the number of donors decreases. They needed fictional body snatchers to keep the system going.

Niven's vision has come true. In 2016, legally donated human organs met less than 10 percent of world need, according to a report by the World Health Organization. In 2014 in the United States alone, the National Kidney Foundation reported that 4,761 Americans died waiting for a transplanted kidney, with a further 3,668 dropping off the list as they became too sick to receive one. It's hardly surprising that some people turn to the black market to save their own lives. Estimates suggest that the illegal trade of all human body organs amounts to between $840 million and $1.7 billion annually, a figure that accounts for roughly 10 percent of transplanted organs. In a throwback to the dark days of Burke and Hare, these surgeries are often done by inadequately trained surgeons under unhygienic conditions.

The question of body piracy is in part taken up in the science-fiction story *Spares* by Michael Marshall Smith. Written in 1996, the book foresees a future where "farms" of clones known as "spares" are kept as an insurance policy of the rich and powerful. Lose an eye, limb, or organ? No problem. Money talks. Your body double is mutilated, and you get your replacement part. Hanging around on the farm to suffer such potential mutilation is the sole point of life for the "spares."

The future foreseen in *Spares* is very similar to the 2005 movie *The Island*. It's 2019. Most of the outside world has been contaminated, or so it seems. A community of people, rescued from the toxic environment, believe they are living in a perfect, but isolated colony. The reality is sinister, however. The colonists are actually clones. They are walking and talking spares, and their sole purpose is to provide medical insurance for their celebrity sponsors. In the face of future horror such as this, maybe someone should listen to Stephen King, who said in an interview with *Time* magazine two years after *The Island*, "I think there ought to be some serious discussion by smart people, really smart people, about whether . . . the whole celebrity culture is healthy. . . . I mean, I know people who can tell you who won the last four seasons on *American Idol* and they don't know who their f***ing Congressmen are."

BLADE RUNNER 2049:

WHEN WILL WE ENGINEER HUMAN LOOK-ALIKES?

Title Card: "Replicants are bioengineered humans, designed by Tyrell Corporation for use off-world. Their enhanced strength made them ideal slave labor. After a series of violent rebellions, their manufacture became prohibited and Tyrell Corp went bankrupt. The collapse of ecosystems in the mid 2020s led to the rise of industrialist Niander Wallace, whose mastery of synthetic farming averted famine. Wallace acquired the remains of Tyrell Corp and created a new line of replicants who obey. Many older model replicants—Nexus 8s with open-ended lifespans—survived. They are hunted down and 'retired.' Those that hunt them still go by the name . . . Blade Runner."

—Michael Green and Hampton Fancher,
screenplay *Blade Runner 2049* (2017)

"The term 'android', which means 'man-like', was not commonly used in science fiction until the 1940s. . . . The word was initially used of automata, and the form 'androides' first appeared in English in 1727 in reference to supposed attempts by the alchemist Albertus Magnus to create an artificial man."

—John Clute and Peter Nicholls,
The Encyclopaedia of Science Fiction (1979)

"The term 'cyborg' is a contraction of 'cybernetic organism' and refers to the product of human/machine hybridization. . . . Elementary medical cyborgs, people with prosthetic limbs or pacemakers . . . have been extrapolated in fiction. There are two other common classes of cyborg in science fiction: functional

cyborgs are people modified mechanically to perform spe-
cific tasks, usually a job of work; adaptive cyborgs are people
redesigned to operate in an alien environment, sometimes so
completely that their humanity becomes problematic."
—John Clute and Peter Nicholls,
The Encyclopaedia of Science Fiction (1979)

ROY BATTY: "I've seen things you people wouldn't believe. Attack
ships on fire off the shoulder of Orion. I watched C-beams glitter
in the dark near the Tannhauser gate. All those moments will be
lost in time . . . like tears in rain. . . . Time to die."
—David Peoples and Hampton Fancher,
screenplay *Blade Runner* (1982)

SEX ROBOTS

The end of the year 2017 saw yet another vision from science fiction starting
to turn into reality—sex robots. Advances in computer tech and engineering design have finally made possible such animatronic sex machines. The
sex tech industry is worth $30 billion a year, and in a recent survey, over
60 percent of heterosexual men admitted they would consider buying a sex
robot. Now the race is on to make the science fiction a more sophisticated
reality. Companies want to profit, surveys say there's huge demand, but
critics say sex robots are dangerous. It's a science-fiction scenario made
flesh. It highlights, once more, the way in which robots, and tech in general,
affects the way in which humans interact with one another.

Such a situation has been imagined in science fiction for decades, of
course. But, before launching into a potted history of human look-alikes,
it's worth spending a few moments clearing up the confusing terminology.
A cyborg and an android are two quite different things. The replicants in
Blade Runner and *Blade Runner 2049* are quite clearly androids, entirely
artificial constructs made to look like a human. The same goes for the
android antagonist, Ash, in the 1979 movie *Alien*, as well as David, the
maintenance man-cum-butler in the 2012 movie *Prometheus* who is
designed to be indistinguishable from humans and begins to develop
his own ego, jealousy, and envy.

Some famous science-fiction icons have been called cyborg when they're actually androids. Take the Arnold Schwarzenegger character in the 1984 movie *Terminator*, for example. Even though the Terminator is dubbed a cyborg assassin sent back in time from 2029, he's clearly an android. (Actually, *Terminator 2* has an even cooler android assassin. The Terminator T-1000 is a liquid-metal shape-shifter. He can easily absorb collateral damage, melt through narrow gaps, and steal through prison bars. He can fashion his hands and arms into any shape he wishes to yield, promising untold possibilities in kitchen and bedroom alike.)

You know the drill. Cyborg is a portmanteau word that is a combination of cybernetics and organism. Organism means any living thing, whereas cybernetics is the scientific study of how humans (or aliens, animals, and machines) control and communicate data. Cyborgs often start off as humans. Consider *Robocop*, for instance. He started off human, then got tooled up to the max to fight future crime. A cyborg can be someone with a heart pacemaker, an artificial joint, or a cochlear implant. Some people think of a cyborg as anyone who feels tech is a part of their biology, any kind of fusion at the "man-machine" interface.

The Tin Man in *The Wizard of Oz* was a cyborg. Many movie-watchers at the time of first seeing L. Frank Baum's 1900 classic brought to the silver-screen in 1939 might not have known it, but the Tin Man started life as a lumberjack named Nick Chopper. (There's a hypothesis, known as nominative determinism, that people tend to gravitate toward areas of work that best fit their names, such as Usain Bolt, Olympic champion and world record-holding sprinter; a Sun Prairie firefighter named Lieutenant Les McBurney; and an Ottawa man charged with indecent exposure who went by the name of Donald Popadick.) We have them in science fiction, too, as Luke Skywalker isn't a bad name for a space pilot. Anyhow, the Tin Man was Dorothy's hero and faithful companion on the Yellow Brick Road. Unfortunately for Nick, those who live by the axe get whacked by the axe. The so-called Wicked Witch of the East conjures up a magic axe that chops off his limbs one by one (the Wicked Witch herself displaying signs of nominative determinism). Like *Robocop* decades later, Nick gradually gets body parts replaced with tech and, lo and behold, becomes a cyborg.

Cyborgs were made famous in America in a novel titled, unsurprisingly, *Cyborg*. Written in 1972 by Martin Caidan, the story is mostly known for the name of its hero, Steve Austin, a test pilot who had a close call with death in an air crash and has large parts of his body replaced with bionic limbs. For most of the mid-1970s, Steve Austin was famous on American TV as *The Six Million Dollar Man*, a reference to how much it cost to rebuild him into a cyborg.

So, when will we see replicants, like those featured in the *Blade Runner* films?

The idea of artificial humans is ancient, dating back to the Golem of Jewish mythology as well as alchemical homunculi, small humans artificially brought into existence through alchemy. Until the nineteenth century, though, it was widely thought that organic compounds would never be synthesized. Humanoid creatures of flesh and blood would have to be created either by magical means or, as in the *Frankenstein* film, by electricity. There appears to have been an imaginative resistance to the idea of the android. Perhaps science-fiction writers dabbled in cyborgs, as it seemed less of a breach of divine prerogative than the building of humanoid automata.

Whatever the history in fiction, for some, the story has now started in fact. Hiroshi Ishiguro is director of the Intelligent Robotics Laboratory at Osaka University in Japan. Ishiguro is sometimes known as the bad boy of Japanese robotics. Comparisons between Ishiguro and Eldon Tyrell of the Tyrell Corporation in *Blade Runner* are very tempting, but we shall resist them. His lab (Ishiguro's, not Tyrell's!) develops androids with lifelike appearance and visible behavior such as facial movements. Ishiguro was quoted by *The Guardian* in 2017 as saying, "I wanted to be a one of the creators to change our life. I want to change this world with robot technologies. . . . I really think I'm doing a very similar thing as the artistic work. I try to represent humanity on robots. So that is my android project."

"Erica" is a case in point. Erica is twenty-three. She is described as beautiful, with a neutral face and a synthesized voice. Even though Erica has twenty degrees of freedom, she can't yet move her hands. Ishiguro is her father and believes that together they will redefine what it means to be human and reveal that the future is closer than we think.

X-MEN:

WILL THERE BE MUTANT HUMANS IN THE FUTURE?

"In some of the most highly developed crustaceans, the whole alimentary canal has solidified into a useless cord, because the animal is nourished by the food in which it swims. The man of the year million will not be bothered with servants handing him things on plates which he will chew, and swallow, and digest. He will bathe in amber liquid which will be pure food, no waste matter assimilated through the pores of the skin. The mouth will shrink to a rosebud thing; the teeth will disappear; the nose will disappear—it is not nearly as big now as it was in savage days—the ears will go away. They are already folded up from what they were, and only a little tip fast vanishing remains to show that ages ago they were long-pointed things which bent forward and backward to catch the sound of approaching enemies."
—H.G. Wells, *The Man of the Year Million* in *The World* (1893)

"When one contemplates the streak of insanity running through human history, it appears highly probable that homo sapiens is a biological freak, the result of some remarkable mistake in the evolutionary process. The ancient doctrine of original sin, variants of which occur independently in the mythologies of diverse cultures, could be a reflection of man's awareness of his own inadequacy, of the intuitive hunch that somewhere along the line of his ascent something has gone wrong."
—Arthur Koestler, *The Ghost in the Machine* (1967)

"The time scale for evolutionary or genetic change is very long. A characteristic period for the emergence of one advanced species

194

from another is perhaps a hundred thousand years; and very often the difference in behavior between closely related species—say, lions and tigers— do not seem very great. . . . But today we do not have ten million years to wait for the next advance. We live in a time when our world is changing at an unprecedented rate. While the changes are largely of our own making, they cannot be ignored. We must adjust and adapt and control, or we perish."
—Carl Sagan, *The Dragons of Eden: Speculations on the Evolution of Human Intelligence* (1977)

"The likelihood is that, in 100,000 years' time, we shall either have reverted to wild barbarism, or else civilization will have advanced beyond all recognition—into colonies in outer space, for instance. In either case, evolutionary extrapolations from present conditions are likely to be highly misleading. . . . Evolution never looks to the future."
—Richard Dawkins, *The Evolutionary Future of Man* (1993)

MUTANTS

Here's a weird story about Charles Darwin you may have never heard. One day, Darwin chanced upon a report of a family of hairy humans. We're not talking *slightly* hairy here. We're talking about some unfortunate souls whose bodies were entirely covered in hair, head to furry foot. So hairy, in fact, that they could comb their faces. They doled out a fortune in shampoo and spent endless hours grooming. Curiosity about these creatures and their rather unusual look was such that, for four generations, they had been kept for others' amusement at the royal court in Burma. Darwin puzzled over these extrahairy apes. He wondered about the quality in these people that, unlike all those around them, made them keep on being hairy, from generation to generation.

Evolutionary history has gifted biologists other fine examples of difference and otherness in human form. Conjoined twins, for example—identical twins that are born fused together; or those affected by cyclopia—being born with one eye, like Homer's famous Cyclops from ancient times; or those suffering from sirenomelia—where the lower limbs are fused together, also known as "mermaid's syndrome." Mutants, one and all. Darwin helped discover that mutation lies at the heart of nature. Without mutant forms, none of the plants and animals on planet Earth would exist in the way they do. And none of them would have been able to develop and evolve in the way they did, through the generations and over the eons. In the *X-Men* franchise, mutants are often too easily dismissed as monsters. The accusers conveniently forget that we are all mutants.

Exactly how we may mutate in the future is one of science fiction's keenest obsessions. In his book *The Time Machine*, H.G. Wells thought that the difference between social classes in Britain would become so great that the classes would eventually evolve into different human species, each species mutant in its own way. One of Wells's future forms of human was the underground-dwelling mutants known as the Morlocks. Wells described them as being "ape-like," with little or no clothing, large eyes, and gray fur covering their bodies. Whether the Morlocks were so hairy they could comb their faces is not known, but they were surely one of science fiction's very first mutants.

A more recent mutant is the Hulk, often called a "strong mutant," as it usually becomes mutated in some accelerated way, rather than the usual gradual and boring kind of evolution. Hulk, of course, was exposed to gamma radiation. In the case of members of the X-Men, they derive their special powers due to their bodies having an "X-gene," which normal humans don't have, and which gifts them superpower status. In some cases, it also seems that the mutations happened because of exposure to radiation.

One of the subtlest mutant forms, in some ways at least, are the Daleks in *Doctor Who*. You might not at first know it, as the Daleks are hidden within their tank-like robotic shells and are actually cyborgs, yet the squidgy alien creatures you seldom see inside the robot shells are the Kaleds, and Kaleds are mutants. The Kaleds were a race of extraterrestrials from the planet Skaro who were deliberately made into strong mutants by their villainous scientist Davros. His final modification, or mutation, was to take away their ability to feel pity for other creatures. Yes, that's right, he deliberately turned them into the uncaring monsters they are (though you would have thought, with all their high-powered tech, they'd be able to do something about that awful voice box). The Daleks are on the dark side of mutation. They soon came to see themselves as the supreme race in the universe, seeking to dominate or exterminate "lesser" races, as if the mutant Daleks were a race of cyborg Nazis.

How mutant humans will turn out is another matter. We have an expectation of our mutant future, thanks to all this science-fiction influence of superheroes and X-Men. And yet some scientists think human evolution is over. Professor Steve Jones of University College London (UCL) has argued that human evolution has reached utopia, or as close to it as we are likely to get. Professor Jones's theory, aptly delivered in the Darwin Lecture Theatre in the Darwin Building of UCL, will disappoint millions of science-fiction fans. Jones bases his prediction on the three component parts that make up evolution: natural selection, random change, and mutation.

As far as natural selection is concerned, Professor Jones says, "In ancient times half our children would have died by the age of twenty. Now, in the Western world, 98 percent of them are surviving to the age of twenty-one. Our life expectancy is now so good that eliminating all accidents and infectious diseases would only raise it by a further two years. Natural selection no longer has death as a handy tool."

With regard to random change, Professor Jones believes such chance is now limited: "Randomness is [an] . . . often forgotten, important ingredient in evolution. Humans are 10,000 times more common than we should be, according to the rules of the animal kingdom, and we have agriculture to thank for that. Without farming, the world population would probably have reached half a million by now—about the size of the population of Glasgow. Small populations which are isolated can change—evolve— at random as genes are accidentally lost. Worldwide, all populations are becoming connected and the opportunity for random change is dwindling. History is made in bed, but nowadays the beds are getting closer together. Almost everywhere, inbreeding is becoming less common. In Britain, one marriage in fifty or so is between members of a different ethnic group, and the country is one of the most sexually open in the world. We are mixing into a global mass, and the future is brown."

And finally, there's evolution through mutation. Are X-Men super-powers on the horizon? Professor Jones suggests not: "Mutation, too, is slowing down. Yes, there are chemicals and radioactive pollution—but one of the most important mutagens is old men. For a twenty-nine-year-old father (the mean age of reproduction in the West) there are around 300 divisions between the sperm that made him and the one he passes on—each one with an opportunity to make mistakes. For a fifty-year-old father, the figure is well over a thousand. A drop in the number of older fathers will thus have a major effect on the rate of mutation. Perhaps surprisingly, the age of reproduction has gone down—the mean age of male reproduction means that most conceive no children after the age of thirty-five. Fewer older fathers means that if anything, mutation is going down. So, if you are worried about what utopia is going to be like, don't—at least in the developed world, and at least for the time being, you are living in it now."

BRAVE NEW WORLD:

WILL EUGENICS BE USED ON HUMAN POPULATIONS IN THE FUTURE?

"One believes things because one has been conditioned to believe them. . . . If one's different, one's bound to be lonely."
—Aldous Huxley, *Brave New World* (1932)

"In a properly organized society like ours, nobody has any opportunities for being noble or heroic. Conditions have got to be thoroughly unstable before the occasion can arise. When there are wars, where there are divided allegiances, where there are temptations to be resisted, objects of love to be fought for or defended—there, obviously, nobility and heroism have some sense. But there aren't any wars nowadays. The greatest care is taken to prevent you from loving anyone too much. There's no such thing as a divided allegiance; you're so conditioned that you can't help doing what you ought to do. And what you ought to do is on the whole so pleasant, so many of the natural impulses are allowed free play, that there really aren't any temptations to resist. And if ever, by some unlucky chance, anything unpleasant should somehow happen, why, there's always soma to give you a holiday from the facts. And there's always soma to calm your anger, to reconcile you to your enemies, to make you patient and long-suffering. In the past you could only accomplish these things by making a great effort and after years of hard moral training. Now, you swallow two or three half-gram tablets, and there you are. Anybody can be virtuous now. You can carry at least half your mortality about in a bottle. Christianity without tears—that's what soma is."
—Aldous Huxley, *Brave New World* (1932)

"But I don't want comfort. I want God, I want poetry, I want real danger, I want freedom, I want goodness, I want sin."

"In fact," said Mustapha Mond, "you're claiming the right to be unhappy."

"All right then," said the Savage defiantly. "I'm claiming the right to be unhappy."

"Not to mention the right to grow old and ugly and impotent; the right to have syphilis and cancer; the right to have too little to eat; the right to be lousy; the right to live in constant apprehension of what may happen tomorrow; the right to catch typhoid; the right to be tortured by unspeakable pains of every kind."

There was a long silence.

"I claim them all," said the Savage at last.

Mustapha Mond shrugged his shoulders. "You're welcome," he said.

—Aldous Huxley, *Brave New World* (1932)

"Did you eat something that didn't agree with you?" asked Bernard. The Savage nodded. "I ate civilization."

—Aldous Huxley, *Brave New World* (1932)

BRAVE NEW WORLD

How do we improve ourselves? The *New York Times* best-sellers list is chockablock with books offering advice. Titles include a book extolling the virtues of the subtle art of not giving a f*ck. If this were really true, surely, they'd be nihilistic enough to actually print the word *fuck* on the cover. Perhaps you should read the next book on the list, a title telling you how badass you are, which is again surely self-defeating, as one finds it hard to imagine a Hitler or a Genghis Khan needing to read such a self-help guide in the first place. Handily, the best-seller list offers an antidote to all this with a "don't bullsh*t yourself!" book.

All this nuance is unnecessary. Science fiction has a history of writers who think that the best way to improve ourselves as a human race is eugenics. The "eu" prefix is Greek and means "good," and the "gen," as in genus, refers to "birth" or "race." Together, eugenics suggests a science of improving the quality of the human population. The word was coined in 1883 by Charles Darwin's cousin, Francis Galton. The coinage is an analogy of ethics, physics, and so on. As Galton put it in his book *Inquiries into Human Faculty and Its Development*, "The investigation of human eugenics, that is, of the conditions under which men of a high type are produced"; this was originally all about improving the human race through bloodlines. People like Dalton argued against the mixing of races, which is known as miscegenation. The *Harry Potter* franchise is one of fiction's best examples. The Death Eater wizards believe they are superior to all other races and argue against miscegenation, using the insult "mudbloods" for those with mixed blood.

Eugenics has a prehistory. Once people realized humans inherited traits from their ancestors, they began to come up with ideas to "improve" the human race. In his book *The Republic*, Greek philosopher Plato argued that an acceptable way to improve the human race was to kill inferior babies at birth. Hardly a vote winner, and so Plato was opposed by another Greek philosopher, Hippocrates, the founder of medicine, whose Hippocratic Oath is still taken by doctors today.

An early science-fiction book, *Gulliver's Travels*, was the first to explore what a society based on eugenics might look like. The novel was written in 1726 by Irish satirist Jonathan Swift. At one point in the story, Gulliver arrives in the land of the Houyhnhnms. These creatures, identical to horses, run a eugenics program involving the selective breeding of their human slaves, known as "yahoos." At first, Gulliver is mistaken for one of the yahoos. But he manages to convince the horse masters that he is intelligent enough to be saved. If he hadn't spoken up, Gulliver would have been sacrificed as part of their cruel eugenics program. One and a half centuries after *Gulliver's Travels* was written, Francis Galton began writing about eugenics. Galton believed in the inequality of humans. For example, he thought Africans were inferior and suggested that the east coast of Africa be settled by the Chinese, who were, according to Galton, superior.

Curiously, a relation of Galton's was to write science fiction's most famous work on an imagined future based on eugenics: Aldous Huxley's *Brave New World*. Written in 1932, Huxley's story takes place about five hundred years into the future, around the year 2540 AD. This Brave New World state is planned at every level. The population has been set at two billion. There are only ten thousand surnames among the entire population, and everything is organized. Through biological engineering and eugenics, the world is extremely conformist, and the population kept quiet through society's vapid consumerist culture.

People are created in hatcheries and bottling plants in which human embryos are developed and conditioned. The human beings created are separated into predetermined classes of alphas, betas, gammas, deltas, and epsilons, based on their intelligence and labor. All are born on the same day, looking the same, acting the same, and being prepared to do the same things. Natural reproduction has been abolished, and the very idea of having a biological mother is considered an obscenity. Outside of this organized society are savage reservations, where life is still left to random chance and is portrayed in a rather bleak way. It's just a different kind of dystopia. This society's efficiency of production minimizes discontent and rebellion. Stability is achieved not just through eugenics, but also through the use of social conditioning. Everyone fits their social role, without dissent.

In the real world, eugenics was in the air in the late nineteenth and early twentieth centuries, along with the argument that the state should take a major role in orchestrating reproduction. As a result, Huxley was caught in a kind of crossfire, finding himself in an ambivalent place with respect to eugenics and social Darwinism. Huxley once said any form of order is better than chaos, clearly indicating that Huxley was torn between social order and individual freedom. Huxley was later to regret his earlier views after the rise of the Third Reich in Nazi Germany, and their own eugenics program.

At a *Brave New World* vs *1984* event in London late in 2017, there was a debate over whether the greatest dystopian novel of the twentieth century was Huxley's *Brave New World* or Orwell's *1984*. British author Will Self spoke in favor of the genius of Huxley's vision of a plastic, engineered technosociety where entertainment is always light and consumerism rampant. In our world today, Self said, pills keep people happy, and virtual reality TV shows distract the masses from actual reality. Self also suggested that Huxley's reality was coming true, as we may soon witness the birth of a new genetic superclass, with gene editing enabling Silicon Valley's superrich to extend their lifespans and enhance the looks and intelligence of their offspring.

Will Self addressed the audience and teased them into confronting their own place in today's brave new world: "I hope everybody's got their phone on airport, or whatever it is, mode because if you didn't have it on airport mode you might receive a little electric shock during the event. Recent studies in cognitive science establish that you get a little jolt of dopamine every time you get an alert on your mobile phone. Every time you push a button on it, or a consumer or a computer console, you are rewarded; there is a pleasurable sensation. You will have noticed in the flow of your screen-mediated lives that there are a myriad of these little electric shocks going on throughout your day. The only difference between you and the children in *Brave New World*, who are conditioned to hate flowers and beautiful things and books, is that your conditioning is happening while you're wide awake. Or are you? Is the current lifestyle of the consumer in fact a kind of waking dream? What Huxley understands only too well is the conditions under what we might call late capitalism. In other words, the kind of neoliberal capitalist societies we live in now. What Huxley understood only too well was that in an economy that is defined by consumption and advertising is the form of behavioral conditioning, everybody will be perfectly pacific as long as their needs and their wants are conflated in their own minds. That is very much the world we're living in. You don't need to feel that you've been conditioned to be conditioned. I think that's the real genius of the dystopic future that Huxley summons up in *Brave New World*."

ORPHAN BLACK:

WHAT DOES THE FUTURE HOLD FOR HUMAN CLONING?

"We live in a science fictional world with things like cloning and face transplants, and things seem to be getting stranger and stranger."
—Alastair Reynolds, *Wales Online* (July 12th, 2009)

"Critics had embraced the somewhat obscure freshman season of *Orphan Black*—a show about identical orphans across the world—but, more than the show itself, had rallied behind its amazing star, Tatiana Maslany, who plays numerous roles as clone variations of the same person (or at least that person's DNA). Often critics are the first to champion low-profile but high-quality shows . . ."
—Tim Goodman, *Hollywood Reporter* (July 25th, 2013)

"That's all it is, there's nothing new about it, identical twins are clones. Anybody who objects to cloning on principle has to answer to all the identical twins in the world who might be insulted by the thought that there is something offensive about their very existence. Clones are simply identical twins. The only really deep reason people have for objecting to such a thing is that it just offends some deep-seated sense in people–what has been called the 'yuck' reaction. It's irrational."
—Richard Dawkins,
BBC World Service interview (January 30th, 1999)

"The cloning of humans is on most of the lists of things to worry about from science, along with behavior control, genetic engineering, transplanted heads, computer poetry, and the unrestrained growth of plastic flowers."

—Lewis Thomas, *A Long Line of Cells: Collected Essays* (1990)

COPIES, COPIES

History is rich with the concept of copies, from the ancient world of Greek philosophy to the likenesses of us that we upload onto modern video games. Since ancient times, we've been mulling over the idea of simulations. Plato believed our physical world was just a copy, a mere reflection of the world of ideas, which was the true reality. For Plato, our visible world of particular things is just a shifting exhibition, like shadows cast on a wall. Another ancient idea, and an equally unsettling one, is that of the Doppelgänger. This is the idea of a spirit, a copy, that looks just like us. It sounds harmless enough, but the idea was you usually saw this copy just before your death! Mary Shelley, the author of *Frankenstein*, wrote that her husband, the Romantic poet Percy Shelley, had seen his Doppelgänger two weeks before his death by drowning.

Scholars say that clones have existed for a long time. As British biologist Richard Dawkins has pointed out, identical human twins are clones. They're made when a single fertilized egg divides into two separate embryos, but while clones are exact copies, twins are not. Their genes are totally the same, but their appearances and their characters differ. The founders of Rome in 770 BC are said to be the twins Romulus and Remus. Romulus is rumored to have killed his brother, an action that may have led to fiction's fascination with the "evil twin."

The idea of evil twins became popular in early movies. One example was a movie made by the famous British comic actor Charlie Chaplin. Based on the idea that satire is the best way to undermine a fascist, and invoking the idea of comedy as protest, Chaplin set about poking fun at Adolf Hitler. In the 1940 anti-Nazi movie, *The Great Dictator*, Chaplin plays a poor Jewish barber who is a dead ringer (we might even say Doppelgänger) for Adenoid Hynckel, a hilarious caricature of Hitler. Disturbingly, censorship and distribution concerns arose during filming. As Chaplin wrote in his 1964 autobiography, "Half-way through making *The Great Dictator* I began receiving alarming messages from United

Artists. They had been advised by the Hays Office that I would run into censorship trouble. Also, the English office was very concerned about an anti-Hitler picture and doubted whether it could be shown in Britain. But I was determined to go ahead, for Hitler must be laughed at."

Once science fiction had begun to explore the idea of human cloning, movies about clones rather than copies or Doppelgängers became more common. The 1978 film *The Boys from Brazil*, about a mad doctor's attempt to clone Hitler and set up a fourth Reich, is a good example.

Cloning is coming. Scientists got to work on cloning animals first. Only a few years before an animal clone was made, US author Michael Crichton had written *Jurassic Park*. Like Mary Shelley's *Frankenstein* before it, *Jurassic Park* was a warning not to mess around with nature. Written in 1990 with the famous movie coming three years later, *Jurassic Park* is about dinosaurs, of course, but the famous theme park monsters were cloned dinosaurs that cause all kinds of unforeseen chaos.

Nonetheless, on July 5th, 1996, the first-ever artificially cloned mammal was born. Dolly the sheep was made of DNA taken from a grown-up female sheep's teat. The breed they used was a Finn Dorset sheep, but the name Dolly was taken after the famous American country and western singer, Dolly Parton (who is also partly famous for her teats). Cloning animals may help preserve endangered species and may also help researching the cloning of human tissue.

Yet the question remains: if human cloning happens in the future, would it benefit everyone, or only those rich enough to afford it? The 2010 movie *Never Let Me Go* is an alternate history, one in which human clones are created to provide organs for naturally born humans, despite the fact that the clones are living, breathing humans themselves. Clones have also made an important appearance in the *Star Wars* universe, of course. *The Clone Wars* was the first weekly television series from Lucasfilm Animation, and it chronicled an adventure that featured the use of clones to quickly make a well-trained and disposable army, which are more adaptive than the droids used by the opposing army.

Finally, on the question of a future of human cloning, it's worth considering the *Four Laws of Behavioural Genetics*, and the way in which the laws impact identical twins. *Law One* says that all traits are partially hereditary. Identical twins reared apart are more similar than fraternal twins reared apart. *Law Two* states that the effect of the genes is larger than the effect of the shared environment. By the time they're adults, identical twins reared together are no more similar than identical twins reared apart. *Law Three* says that a lot of variance in behavioral traits isn't down to either genes or the shared environment, which means that identical twins reared together aren't truly identical. *Law Four* states that complex traits are usually shaped by many genes of small effect. For example, there's no common genes that affect IQ by, say, five points, but there are thousands of genes that effect IQ by a tiny fraction of a point. As Mary Shelley once said, technology is a two-edged sword. It can be used for good, and bad. It's clear from the *Four Laws of Behavioural Genetics* that cloning will have consequences hard to predict, which means that science fiction will have plenty to write about for decades to come.

BLACK MIRROR AND ALTERED CARBON:

WHEN WILL HUMAN CONSCIOUSNESS SLIP ITS SKIN?

"*Altered Carbon* is set in a future where consciousness is digitized and stored in cortical stacks implanted in the spine, allowing humans to survive physical death by having their memories and consciousness 're-sleeved' into new bodies."

—*Altered Carbon* television series, *Netflix* (2018)

"We have no idea, now, of who or what the inhabitants of our future might be. In that sense, we have no future. Not in the sense that our grandparents had a future or thought they did. Fully imagined cultural futures were the luxury of another day, one in which 'now' was of some greater duration. For us, of course, things can change so abruptly, so violently, so profoundly, that futures like our grandparents' have insufficient 'now' to stand on. We have no future because our present is too volatile. We have only risk management. The spinning of the given moment's scenarios. Pattern recognition."

—William Gibson, *Pattern Recognition* (2003)

"Another example—maybe a better one, in a way—was when it was confirmed that Michael Jackson was going to marry Elvis Presley's daughter. [loud, game-show buzzer noise] A good friend of mine in the States faxed me, and he simply . . . he said, 'This makes your job more difficult.' And I knew exactly, I knew exactly what he meant. 'Cos something . . . a scenario that seemed to belong to the universe of the late Terry Southern, was suddenly, suddenly real. It's that truth-is-stranger-than-fiction factor keeps getting jacked up on us on a fairly regular, maybe even exponential, basis. I think that's something peculiar to our time. I don't think our grandparents had to live with that."

—William Gibson, *No Maps for These Territories* (2000)

MARTHA: "Jump."

ASH-2: "What? Over there? I never expressed suicidal thoughts, or self-harm."

MARTHA: "Yeah, well, you aren't you, are you?"

ASH-2: "That's another difficult one, to be honest with you."

MARTHA: "You're just a few ripples of you. There's no history to you. You're just a performance of stuff that he performed without thinking, and it's not enough."

—*Black Mirror* television series, episode 2.01 written by Charlie Brooker (2013)

BLACK MIRROR

The American entertainment company *Netflix* is making wonderful science-fictional waves early in the twenty-first century. We've had original and commissioned series such as *Stranger Things* and *Star Trek: Discovery*, superhero series including *Luke Cage*, *Daredevil*, and *Jessica Jones*, as well as a remake of the 1960s classic television series *Lost in Space* (*"Danger, Will Robinson. Danger! Danger!"*). Then there's two excellent series looking at the social consequences of new tech: *Black Mirror* and *Altered Carbon*.

Black Mirror is a British science-fiction anthology created by English humorist and author Charlie Brooker. Winner of Emmy and BAFTA awards, the series met with critical acclaim and became renowned for making predictions that later came true. Foremost of these was an episode called *The National Anthem*. The story revolves around a British prime minister being blackmailed into having sex with a pig. In September 2015, four years after the episode was broadcast, it was revealed that David Cameron, who at the time was British prime minister, had placed his "private part" into the mouth of a dead pig as part of a university initiation rite. In an interview with *The Guardian*, Brooker referred to the "pig-gate" incident as a "coincidence, albeit a quite bizarre one," saying he was quite perturbed when he first heard the story. "I did genuinely for a moment wonder if reality was a simulation, whether it exists only to trick me."

The second season of *Black Mirror* opened with an episode called *Be Right Back*. The tale is about a social media obsessive named Ash who lives with his girlfriend Martha. Ash, known for spending hours at a time on social media, dies suddenly in a real-life traffic accident. Days after Ash's death, Martha discovers she's pregnant and creates a version of Ash using new technology, which is able to simulate his personality and voice on her smartphone based on Ash's digital identity, including his social media profile. At first, the uploaded version of Ash helps Martha overcome her despair, but soon enough the artificial Ash (a passing reference to the science officer in Ridley Scott's *Alien*?) tells her of a new experimental stage, and she agrees to have his replica uploaded into a synthetic android body, almost identical to Ash's. The android fails to meet Martha's expectations, however, as it's not able to reproduce the nuances and small human idiosyncrasies she associated with her loved one's behavior. Martha starts to distance herself from the android until, ultimately, Martha takes the artificial Ash to a cliff and orders it to jump off. At first, the android starts to obey her instruction, and Martha shows her frustration that Ash would never have simply obeyed. In reply, the android begs for its life, and Martha finally agrees to store the android in the attic, where, a few years later, her little daughter visits it every weekend.

ALTERED CARBON

The idea of human consciousness surviving death is taken one step further in Richard Morgan's 2002 science-fiction novel *Altered Carbon*. It's a future in which interstellar travel is made possible by transferring human consciousness between bodies known as "sleeves." Within a body, or sleeve, human minds are stored on "stacks," or discs that live in the back of the neck. Traditional death has been transformed into a mere movement of the mind from one bag of flesh to the next. The Catholics of this future society choose not be "resleeved," as they believe that the soul goes to Heaven when they die.

Most people in *Altered Carbon* can afford to get resleeved at the end of their lives, but as they're unable to update their bodies, most humans age fully each time, which discourages most people from resleeving more than once or twice. So, while humans can live indefinitely in theory, in practice most people choose not to. Only the wealthiest are able to appropriate substitute sleeves on an ongoing basis. For this reason, the wealthy are long-lived and known as Meths, after the Biblical figure Methuselah, the man alleged to have lived to the age of 969 in the Hebrew Bible. The wealthy can also afford to keep copies of their minds in remote storage, which are regularly updated, and ensures that even if their stack is destroyed, they can be resleeved.

This question of uploading human consciousness is also featured in the work of Professor Robin Hanson, economist and scholar at Oxford's Future of Humanity Institute. Professor Hanson imagines taking the best and brightest two hundred scientists, scanning their brains, and uploading their consciousness into androids—robots indistinguishable from the humans upon which they are based, except a thousand times faster and fitter for the future.

If such a future were possible, Professor Hanson concludes, androids will displace humans in almost every walk of life and work. Some androids will have bodies, but other minds will simply live in virtual reality and dispense with bodies completely. The androids will work almost continually but also choose to dream about an existence that is nearly all leisure. Surveillance will be total. And yet, if history is anything to go by, there'll be a popular human revolution long before that happens. Hopefully.

THE AMAZING SPIDER-MAN:

WHY IS THE WORLD HUNG UP ON THE IDEA OF A SUPERSERUM?

NARRATOR: "Steve Rogers! Too puny, too sickly, to be accepted by the army! Steve Rogers! Chosen from hundreds of similar volunteers because of his courage, his intelligence, and his willingness to risk death for his country if the experiment should fail!"

DOCTOR ERKSINE: "If we have erred, Rogers will be dead within seconds! For, he is drinking the strongest chemical potion ever created by man! But, if we succeed, he will be the first of an army of fighting men such as the world has never known! His reflexes, his physical condition, his courage, will be second to none!"

—Stan Lee, *The Origin of Captain America*,
Tales of Suspense #63 (February, 1965)

"The most racking pangs succeeded: a grinding in the bones, deadly nausea, and a horror of the spirit that cannot be exceeded at the hour of birth or death. Then these agonies began swiftly to subside, and I came to myself as if out of a great sickness. There was something strange in my sensations, something indescribably sweet. I felt younger, lighter, happier in body; within I was conscious of a heady recklessness, a current of disordered sensual images running like a millrace in my fancy, a solution of the bonds of obligation, an unknown but innocent freedom of the soul. I knew myself, at the first breath of this new life, to be more wicked, tenfold more wicked, sold a slave to my original evil and the thought, in that moment, braced and delighted me like wine."

—Robert Louis Stevenson,
The Strange Case of Dr. Jekyll and Mr. Hyde (1886)

[Connors has set up a minilab in the sewer and is recording himself on camera]

DR CURT CONNORS: "Subject; Dr. Curtis Connors. Current temperature; 89.7, steady for forty-eight hours. Blood panels reveal lymphocyte and monocyte readings consistent with subject's past. Clotting rate vastly improved, marked enhancement in . . . in muscle response, strength, and elasticity.

[as Connors is testing himself, Peter [Parker], down the sewer, builds a giant web in the intersection of the tunnels; back at his sewer lab, Connors continues talking into the camera]

CONNORS: "Eye sight similarly improved. Subject no longer requires corrective lenses. This is no longer about curing ills. This is about finding perfection. In attempt to regress regenerative relapse, dosage has been increased to two hundred milligrams."

[he injects himself with the serum and suddenly turns violent and knocks the camera out; we then see that Peter has set up a camera in the sewer tunnel and lies in his web to wait for Connors]

—James Vanderbilt, Alvin Sargent, and Steve Kloves,
The Amazing Spider-Man (2012)

DR MELIK: "This morning for breakfast he requested something called "wheat germ, organic honey, and tiger's milk."

DR ARAGON: [chuckling] "Oh, yes. Those are the charmed substances that some years ago were thought to contain life-preserving properties."

DR MELIK: "You mean there was no deep fat? No steak or cream pies or . . . hot fudge?"

DR ARAGON: "Those were thought to be unhealthy . . . precisely the opposite of what we now know to be true."

DR MELIK: "Incredible."

—Woody Allen and Marshall Brickman, *Sleeper* (1973)

SUPERSERUM

It's just a swig away, the monster or hero within us. Take Robert Louis Stevenson's famous 1886 novel, *The Strange Case of Dr. Jekyll and Mr.*

Hyde, for example. With the mere drinking of a draught, the prim and proper society doctor in Jekyll was able to awaken within himself the brutal libertine, Mr. Hyde. Stevenson's suggestion is that civilization is only skin deep. Now, it may be true that ancient Greek myths included the concept of ambrosia, a food or drink of the gods that when supped would give the supper longevity, or immortality. And it's also true that, in medieval Christian church doctrine, "man" sat midway between the inert clay of the Earth's basely core and the divine spirit of heaven. Humans were both body and soul and so possessed the dramatic choice of either following their base and human nature down to its natural place in Hell, or engaging with the spirit and following the soul up through the celestial spheres to God.

But Stevenson's story is really something quite different. It's a science fiction of the age. In the closing decades of the nineteenth century, writers were beginning to explore the influences of science on the human frame. What does the post-Darwinian scientist Jekyll find, once he'd downed his drug? Something of an evil and lustful nature, something gothic at the heart of the human. Stevenson called on Darwin to invoke humanity's beastly core and prefigured Sigmund Freud in showing that the id sometimes overwhelms the ego, even in polite Victorian English society.

Jekyll and Hyde soon became associated with Jack the Ripper. On August 7th, 1888, just three days after a *Dr. Jekyll and Mr. Hyde* stage production of Stevenson's novel opened in London's West End theatre district, a Martha Tabram was found stabbed to death in the Whitechapel neighborhood of the city. At the end of the month, Mary Ann Nichols was found mutilated and murdered, too. These and other Whitechapel murders of local women caused an uproar in London, and the suspicion was that the murders were committed by a lone serial killer, who came to be known as Jack the Ripper. Rumors arose that The Ripper led a respectable life by day and a demonic one by night, just like Jekyll and Hyde. So it was that, on October 5th, 1888, the City of London Police received a letter identifying Jack the Ripper as none other than Richard Mansfield, the lead actor in the *Dr. Jekyll and Mr. Hyde* stage play. The letter writer figured that as Mansfield was so good in his performance as both Jekyll and Hyde, it would take little more application for Mansfield to

MARK BRAKE 🚀 215

play by day and murder disguised and undetected by night. (Incidentally, the stage production's acting manager was none other than Bram Stoker, who would later write the horror novel *Dracula*).

So began another science-fiction infatuation. The long flirtation with drugs included those whose special powers or abilities were gained unnaturally, such as the digestion or injection of elixirs by Doctors Jekyll and Moreau, as well as the Invisible Man and Captain America. Indeed, writers have long imagined a world in which humans were forced to take drugs, a world in which the average person on the street had little choice on whether something was being added to their metabolic systems. Aldous Huxley's 1932 novel, *Brave New World*, is about a regimented and mechanized drug-induced utopian society. The novel's title is a reference to a passage in Shakespeare's *The Tempest*: "O wonder! How many Godly creatures are there here! How beauteous mankind is! O Brave new world! That has such people in it!" It's maybe the only line of Shakespeare to be made famous by someone else, for *Brave New World* is Huxley's dystopia of eugenics, the drug Soma, and microgravity tennis. It's a world organized for optimum social stability, a world, incidentally, upon which literary references to Shakespeare would be entirely lost. Huxley's vision is a stark warning about the widespread consumption of pharmacology and mass medication, and even though the book is a lurid and satirical dystopia about the hopes and fears of the 1930s, it's also uncannily prescient about our own time.

In the 2012 movie, *The Amazing Spider-Man*, the film's very own half-lizard mad doctor Curt Connors plans to make all humans lizard-like by letting loose a superserum. His launchpad of choice for the release of the chemical cloud is the top of Oscorp Tower, and his ultimate aim is to rid humanity of the inherent weaknesses that he believes plague it. Connors's plan is foiled by Spider-Man's last-ditch attempt to disperse an antidote cloud instead, restoring Connors and earlier victims to mere Jekylls once more.

Our infatuation with superserums is lucratively exploited by modern consumer capitalism. The merest brief excursion on a Google search for "superserum" returns copious hits, including: conditioning oils for every nook and knackered cranny of your tired body, skincare products that tackle a multitude of complexion concerns you didn't even know you had, and a "mountain of youth, genius in a bottle, 6-in-1 anti-aging biopeptide wrinkle-smoothing moisture shot." As Bob Dylan once said, "it's easy to see, without looking too far, that not much is really sacred."

If consumers tire of the endless vanity of superserums, they can become similarly obsessed with superfoods—the types of food believed to have life-transforming health benefits. Blueberries are considered a superfruit due to their high concentration of antioxidants and dietary fiber. But, allegedly, the most potent superfood is cacao ("ka-cow"), the raw, uncooked, form of chocolate. Scientific investigation has shown that raw cacao contains over 100 chemical constituents, including amino acids, vitamins, minerals, polyphenols, alkaloids, phospholipids, serotonin, tryptophan, protein, and "much, much more!"